# ALGORITHMS AND TECHNIQUES
## FOR
## VLSI LAYOUT SYNTHESIS

# THE KLUWER INTERNATIONAL SERIES
# IN ENGINEERING AND COMPUTER SCIENCE

## VLSI, COMPUTER ARCHITECTURE AND
## DIGITAL SIGNAL PROCESSING

*Consulting Editor*

Jonathan Allen

**Other books in the series:**

*Logic Minimization Algorithms for VLSI Synthesis.* R.K. Brayton, G.D. Hachtel, C.T. McMullen, and A.L. Sangiovanni-Vincentelli. ISBN 0-89838-164-9.
*Adaptive Filters: Structures, Algorithms, and Applications.* M.L. Honig and D.G. Messerschmitt. ISBN 0-89838-163-0.
*Introduction to VLSI Silicon Devices: Physics, Technology and Characterization.* B. El-Kareh and R.J. Bombard. ISBN 0-89838-210-6.
*Latchup in CMOS Technology: The Problem and Its Cure.* R.R. Troutman. ISBN 0-89838-215-7.
*Digital CMOS Circuit Design.* M. Annaratone. ISBN 0-89838-224-6.
*The Bounding Approach to VLSI Circuit Simulation.* C.A. Zukowski. ISBN 0-89838-176-2.
*Multi-Level Simulation for VLSI Design.* D.D. Hill and D.R. Coelho. ISBN 0-89838-184-3.
*Relaxation Techniques for the Simulation of VLSI Circuits.* J. White and A. Sangiovanni-Vincentelli. ISBN 0-89838-186-X.
*VLSI CAD Tools and Applications.* W. Fichtner and M. Morf, editors. ISBN 0-89838-193-2.
*A VLSI Architecture for Concurrent Data Structures.* W.J. Dally. ISBN 0-89838-235-1.
*Yield Simulation for Integrated Circuits.* D.M.H. Walker. ISBN 0-89838-244-0.
*VLSI Specification, Verification and Synthesis.* G. Birtwistle and P.A. Subrahmanyam. ISBN 0-89838-246-7.
*Fundamentals of Computer-Aided Circuit Simulation.* W.J. McCalla. ISBN 0-89838-248-3.
*Serial Data Computation.* S.G. Smith and P.B. Denyer. ISBN 0-89838-253-X.
*Phonological Parsing in Speech Recognition.* K.W. Church. ISBN 0-89838-250-5.
*Simulated Annealing for VLSI Design.* D.F. Wong, H.W. Leong, and C.L. Liu. ISBN 0-89838-256-4.
*Polycrystalline Silicon for Integrated Circuit Applications.* T. Kamins. ISBN 0-89838-259-9.
*FET Modeling for Circuit Simulation.* D. Divekar. ISBN 0-89838-264-5.
*VLSI Placement and Global Routing Using Simulated Annealing.* C. Sechen. ISBN 0-89838-281-5.
*Adaptive Filters and Equalisers.* B. Mulgrew, C.F.N. Cowan. ISBN 0-89838-285-8.
*Computer-Aided Design and VLSI Device Development, Second Edition.* K.M. Cham, S-Y. Oh, J.L. Moll, K. Lee, P. Vande Voorde, D. Chin. ISBN: 0-89838-277-7.
*Automatic Speech Recognition.* K-F. Lee. ISBN 0-89838-296-3.
*Speech Time-Frequency Representations.* M.D. Riley. ISBN: 0-89838-298-X.
*A Systolic Array Optimizing Compiler.* M.S. Lam. ISBN: 0-89838-300-5.

**AT&T**

# ALGORITHMS AND TECHNIQUES
# FOR
# VLSI LAYOUT SYNTHESIS

by

**Dwight Hill**
**Don Shugard**
**John Fishburn**
**Kurt Keutzer**

AT&T Bell Laboratories

**Kluwer Academic Publishers**
**Boston/Dordrecht/London**

**Distributors for North America:**
Kluwer Academic Publishers
101 Philip Drive
Assinippi Park
Norwell, Massachusetts 02061 USA

**Distributors for the UK and Ireland:**
Kluwer Academic Publishers
Falcon House, Queen Square
Lancaster LA1 1RN, UNITED KINGDOM

**Distributors for all other countries:**
Kluwer Academic Publishers Group
Distribution Centre
Post Office Box 322
3300 AH Dordrecht, THE NETHERLANDS

**Library of Congress Cataloging-in-Publication Data**

Algorithms and techniques for VLSI layout synthesis / by Dwight Hill
... [et al.].
    P.   cm. -- (The Kluwer international series in engineering and
computer science ; 65. VLSI, computer architecture, and digital
signal processing)
  Bibliography: p.
  Includes index.
  ISBN 0-89838-301-3
  1. Integrated circuits--Very large scale integration--Design and
construction--Data processing.  I. Hill, Dwight D.  II. Series:
Kluwer international series in engineering and computer science ;
SECS 65.  III. Series: Kluwer international series in engineering
and computer science. VLSI, computer architecture, and digital
signal processing.
  TK7874.A42 1989
621.381 '73--dc19                         88-7584
                                        CIP

Printed in the United States of America

# Table of Contents

Chapter 1:    **Introduction**........................................................   **1**

Chapter 2:    **The IMAGES Language**.......................................   **5**

    2.1 Background   6

    2.2 Principal Constructs   6

    2.3 Geometric Constraints   8

    2.4 Geometric Constraint Resolution   10

    2.5 Electrical Connectivity   30

    2.6 Other IMAGES Features   31

    2.7 Interface with the Rest of IDA   33

    2.8 Quick Access to Large Designs   33

    2.9 The IMAGES Implementation   36

    2.10 Summary   36

Chapter 3:    **Designer Interaction**.............................................   37

    3.1 Basic Editing Environment   37

    3.2 Hybrid Logic/Layout Diagrams   51

    3.3 Interactive Simulation   55

    3.4 Compaction   60

    3.5 Reading from the Top-Down   61

    3.6 Editor Internals   62

    3.7 Moving on - Icon as a Non-graphical Tool   63

Chapter 4:    **Geometric Algorithms**..........................................   **65**

|  | 4.1 Enhancing the Basic IMAGES | |
|  | Data Structures | 65 |
|  | 4.2 Scan-Line Technique | 66 |
|  | 4.3 Consolidating Rectangles | 68 |
|  | 4.4 Tub Insertion | 69 |
|  | 4.5 Hierarchical Net Extraction | 77 |
|  | 4.6 Hierarchical Design Rule Checking | 92 |
|  | 4.7 Summary | 104 |
| **Chapter 5:** | **Enhancements to the IMAGES Language** | |
|  | **for Synthesis** ................ | **105** |
|  | 5.1 Stretchable Routing using | |
|  | IMAGES Constraints | 105 |
|  | 5.2 Impact on Design Methods | 114 |
|  | 5.3 Designer Written "Generators" | 114 |
|  | 5.4 Design Rule Updatable Generators | 121 |
|  | 5.5 Summary: Advantages and Limitations | 126 |
| **Chapter 6:** | **Automatic Layout of Switch-Level Designs**.......... | **129** |
|  | 6.1 System Organization | 129 |
|  | 6.2 Steps in SC2D | 131 |
|  | 6.3 Steps After SC2D | 138 |
|  | 6.4 Detailed Layout in SC2 | 138 |
|  | 6.5 Performance and Parallel Processing | 149 |
|  | 6.6 Summary | 150 |
| **Chapter 7:** | **Switch-Level Tools**.................... | **153** |
|  | 7.1 Transistor Sizing with TILOS | 153 |
|  | 7.2 SLICC - An Alternative to Schematics | 169 |
| **Chapter 8:** | **Summary and Trends**............................ | **181** |
|  | 8.1 Summary | 181 |
|  | 8.2 VLSI: From Science Fiction to Commodity | 183 |
|  | 8.3 Design Style Trends | 184 |
| **Appendix A:** | **Data Structures** ...................... | **189** |
|  | A.1 Basic IDA Data Structures | 189 |
|  | A.2 Basic Design Elements | 189 |
| **Appendix B:** | **Background on the "i" Language** ...................... | **199** |
| **REFERENCES:** | ...................................................... | **201** |
| **INDEX:** | ...................................................... | **209** |

# List of Figures

| | | |
|---|---|---|
| 2-1 | An IMAGES Leaf Cell. | 7 |
| 2-2a | The Corresponding Layout. | 8 |
| 2-2b | The Level Key. | 8 |
| 2-3 | Showing the Results of Algorithms 2-1a and 2-1b | 15 |
| 2-4 | Showing the Results of Algorithm 2-1c | 17 |
| 2-5 | A Constraint Graph as Solved by Algorithm 2-2a | 20 |
| 2-6 | A More General Constraint Graph as Solved by Algorithms 2-3 and 2-4 | 22 |
| 2-7 | Building $G'$ Using Algorithm 2-5a | 25 |
| 2-8 | Example of the Need for Backward Edges | 27 |
| 2-9 | $G_{bak}$ Associated with Figure 2-8 | 28 |
| 3-1 | Initial Icon Screen Display. | 38 |
| 3-2 | Manipulation of the Chosen Group. | 40 |
| 3-3a | Before Resize Command. | 41 |
| 3-3b | After Resize Command. | 42 |
| 3-4a | Original Layout Borrowed from Another Design. | 43 |
| 3-4b | Layout Pruned Down. | 44 |
| 3-4c | Layout Ready for Replication. | 45 |
| 3-4d | Replicated Layout. | 46 |

| | | |
|---|---|---|
| 3-4e | Flattened Layout. | 46 |
| 3-4f | Touched up. | 47 |
| 3-5 | Schematic for XOR Gate. | 47 |
| 3-6a | Hybrid Layout-Schematic. | 53 |
| 3-6b | Actual Layout. | 54 |
| 3-7 | Relationship Between Editor and Simulator. .. | 56 |
| 3-8a | Schematic Simulation Startup. | 57 |
| 3-8b | Schematic Simulation Running. | 58 |
| 3-9 | Data Structures Supporting Interactive Graphic Simulation. | 59 |
| 3-10 | XOR Gate After Compaction. | 61 |
| 4-1a | The Edges of Two Abutting Wires. | 67 |
| 4-1b | The Sorted List of Edges. | 67 |
| 4-2a | Starting Geometry. | 68 |
| 4-2b | Ending Geometry. | 68 |
| 4-3a | The Corner Overlap Category. | 69 |
| 4-3b | The Crossing Category. | 69 |
| 4-4a | A CMOS Inverter. | 71 |
| 4-4b | Inverter Diffusion Areas Needing a Tub (metal is shown for context). | 72 |
| 4-5a | Tub Contact Positions. | 73 |
| 4-5b | Resulting Tree. | 73 |
| 4-6a | Likely Pull-Apart. | 75 |
| 4-6b | Consolidated Rectangles. No Pull-Apart. | 75 |
| 4-7a | Two Tub Rectangles. | 76 |
| 4-7b | One Large Tub Rectangle. | 77 |
| 4-8 | Three Net Areas of the IMAGES Device. | 79 |
| 4-9 | The Value of the "Connected" Field at each Scan-line Location. | 80 |
| 4-10a | The Linked List Formed by the "Connected" Field. | 81 |
| 4-10b | The Nets After Consolidation. | 81 |
| 4-11a | The Internals of Symbol X. | 83 |
| 4-11b | A View of the Instantiated Symbol X. | 83 |
| 4-12a | Some Fields of the Net Structure. | 85 |
| 4-12b | Symbol X with Its Nets. | 85 |
| 4-12c | The Net Fields After Inst-port "a" Passes Through the Main Loop. | 86 |
| 4-12d | The Net Fields After Inst-port "b" Passes Through the Main Loop. | 86 |

4-12e    The Net Fields After Inst-port "c"
         Passes Through the Main Loop. .................    87
4-12f    The Net Fields After All the Inst-ports
         Are Done. ....................................    87
4-13     A Schematic of a Set-reset Flip Flop. ........    90
4-14     A Virtual-grid Cell with Shorted Poly
         Beneath Port "b". ............................    91
4-15a    The Internals of Symbol "s". .................    94
4-15b    Instance "i1" of Symbol "s" Maps
         Net "b" to Net "g". ..........................    94
4-16     Edge Storage Array Linked by Level. ..........    95
4-17     Symbol x Has Bounding Box Coordinates
         (-1,-1) (3,5). ...............................    96
4-18a    Symbol "e". ..................................    97
4-18b    Instance "m" of Symbol "e" Rotated
         270 Degrees, and a Box. ......................    97
4-18c    Box Rotated into Symbol "e's" Coordinate
         System. ......................................    98
4-19a    Symbol e. ....................................    99
4-19b    Symbol m Containing Instance of Symbol
         e Rotated 270 Degrees. .......................    99
4-19c    High-level Symbol Containing Rotated
         Instance of m and a Box. .....................    99
4-19d    Relationship as seen by Design Rule Checker ...   99
4-20     The Error Box Saved by the "Trfind" Flag. ....   101
4-21a    A Notch Formed Between Features of
         the Same Level. ..............................   103
4-21b    The Patch Inserted to Cover the Notch
         in Figure 4-21a. .............................   103
5-1a     A Simple Route. ..............................   107
5-1b     A More Complex Designer-Entered Route. .......   108
5-2      A More Complex Routing Job. ..................   112
5-3      Relation of Generator Building Tools. ........   116
5-4a     A Fragment of a Simple Generator in IMAGES-C.   117
5-4b     Resulting C Language Output. .................   118
5-4c     Resulting IMAGES Language Output. ............   119
5-4d     Final Layout. ................................   120
5-5      Relation of Compacter in Method 1. ...........   122
5-6      RAM Generator Output Using Method 1. .........   123
5-7      Relation of Compacter in Method 2. ...........   124

| 5-8 | Counter Generator Built using Method 2. ............... | 125 |
| 5-9 | Relation of Compacter in Method 3. ....................... | 126 |
| 5-10 | Potential Problems in Writing and Using A Generator. ...................................................... | 127 |
| 6-1 | System Architecture of SC2D. .............................. | 130 |
| 6-2 | SC2D Layout Comparing Area with and without Intra-row Routing. ....................................... | 137 |
| 6-3a | Cycle Introduced by the SC2 Router. ...................... | 144 |
| 6-3b | Cycle Corrected. ............................................. | 145 |
| 6-4 | Metal2 Substituted for Poly Directly. ...................... | 146 |
| 6-5 | Metal2 Used Horizontally. ................................. | 147 |
| 6-6a | Routing Before Via Elimination. ............................ | 148 |
| 6-6b | Routing After Via Elimination. ............................. | 149 |
| 6-7 | Parallel Processing with SC2D. ............................. | 150 |
| 7-1 | Memory/Combinational-Logic Model of Digital CMOS Circuits. ....................................... | 154 |
| 7-2 | TILOS's Electrical Model of a Transistor. ............... | 155 |
| 7-3 | Delay Through Pulldown Network Modeled with RC Network. ........................................... | 156 |
| 7-4 | Transistor Sizing Used To Reduce Delay Due To Fanin. ................................................ | 160 |
| 7-5 | Transistor Sizing Used To Reduce Delay Due To Fanout. ............................................... | 161 |
| 7-6 | Transistor Sizing Used To Defeat Wire Capacitance. ................................................. | 162 |
| 7-7 | Calculating Sensitivities. .................................... | 164 |
| 7-8 | TILOS/SC2 Versus Standard Cell. ........................ | 167 |
| 7-9a | Result for x = !(a && b). ................................... | 171 |
| 7-9b | Result for  y = !(!(e && !(f ‖ g)). ....................... | 172 |
| 7-10a | Pass Logic with Directive Optimize Area. ............... | 177 |
| 7-10b | Pass Logic with Directive Optimize Speed. ............. | 178 |
| A-1 | Relationship between "World" and User Symbols. | 190 |
| A-2 | The Ring Abbreviation of Figure A-1. ................... | 191 |
| A-3 | Symbol and Its Objects. .................................... | 192 |
| A-4 | Connections and Objects. .................................. | 197 |
| A-5 | A Group Pointing to Some Members. ..................... | 197 |

# List of Algorithms

| | | |
|---|---|---|
| 2-1a | Find the Root of $v$, and Collapse. | 14 |
| 2-1b | Propagating Value to Root. | 14 |
| 2-1c | Union. | 16 |
| 2-1d | Clean_up. | 17 |
| 2-2a | Breadth-First Search. | 19 |
| 2-2b | Forward Propagate. | 19 |
| 2-3 | General Inequality Constraint Solver. | 21 |
| 2-4 | More Efficient General Inequality Solver. | 23 |
| 2-5a | Moving Equivalence Class Members. | 25 |
| 2-5b | Value Update. | 26 |
| 2-5c | Solving Systems of Equalities and Inequalities. | 26 |
| 2-6a | Re-expressing Inequalities. | 28 |
| 2-6b | Solving General Systems of Equalities and Inequalities. | 29 |
| 3-1 | The Procedure Used for Finding Symbolic References. | 51 |
| 3-2 | A Procedure of Commands to the Editor. | 63 |
| 4-1 | Simple Tub Insertion Algorithm. | 70 |
| 4-2 | The Procedure for Searching the Binary Tree. | 74 |
| 4-3 | Gathering and Reclaiming Linked Nets. | 82 |
| 4-4 | Associating Nets with Instance-Ports. | 84 |

6-1 Regular Design Placement. ....................................... 133

6-2 SC2's Transistor Splitting. ...................................... 140

6-3 Algorithm for Ordering FETs to Minimize Breaks. 142

6-4 Simple Via Elimination Heuristics. ........................ 148

7-1 Translating Syntax to FETs ..................................... 170

7-2 Eliminating Redundant Logic Gates ....................... 173

# Acknowledgements

We would like to thank AT&T Bell Labs in general for the support in working on this system, and in allowing this work to be documented and released. In addition, we acknowledge that many other people have worked on IDA and many continue to do so. Specifically Jim Allen, Sanjay Bajekal, Dave Bearden, Bill Bullman, John Carelli, Stu Carpenter, Mike Condict, Leon Davieau, Jalil Fadavi-Ardekani, Bill Fischer, Kalyan Mondal, Howard Moscovitz, Glenn O'Donnell, Dom Petitti, Larry Rigge and Dave Willauer are currently working on the IDA system.

We are especially indebted to Jim Allen for his software contributions, thoughtful criticisms and patience with the problems associated with the evolving system.

The IMAGES language was the creation of a working group consisting of, in addition to the authors, Bryan Ackland, Al Dunlop, Van Kelly, Howard Moscovitz and John Tauke. Mike Condict is the current IMAGES "language engineer" and has made significant contributions to both the language and its implementation. Stan Schuyler and Sanjiv Taneja also share credit for the growing success of the IMAGES language.

Early contributions to IDA were made by Stu Brown, Sally Browning, Misha Buric, Carl Christensen, K. C. Chu, Ed Goldberg, Peter Honeyman, Steve Johnson, Mary Leland, Ed Lien, John Manferdelli, Tom Matheson, Boyd Mathews, Mike Maul and Paul Rubin.

Many suggestions from colleagues in the CAD community and from users of IDA have led to improvement to the quality of the software. Our CAD colleagues from Labs 1127, 1135, 5221, and 5217 have been providing useful comments and feedback. Particular thanks to Duane Aadsen, Lin-Du Chen, John Sharp and Joe Wroblewski in this regard.

# ALGORITHMS AND TECHNIQUES
## FOR
## VLSI LAYOUT SYNTHESIS

# Chapter 1
# Introduction

This book describes a system of VLSI layout tools called IDA which stands for "Integrated Design Aides." It is not a main-line production CAD environment, but neither is it a paper tool. Rather, IDA is an experimental environment that serves to test out CAD ideas in the crucible of real chip design. Many features have been tried in IDA over the years, some successfully, some not. This book will emphasize the former, and attempt to describe the features that have been useful and effective in building real chips.

Before discussing the present state of IDA, it may be helpful to understand how the project got started. Although Bell Labs has traditionally had a large and effective effort in VLSI and CAD, researchers at the Murray Hill facility wanted to study the process of VLSI design independently, emphasizing the idea of small team chip building. So, in 1979 they invited Carver Mead to present his views on MOS chip design, complete with the now famous "lambda" design rules and "tall, thin designers." To support this course, Steve Johnson (better known for YACC and the portable C compiler) and Sally Browning invented the constraint-based "i" language and wrote a compiler for it. A small collection of layout tools developed rapidly around this compiler, including design rule checkers, editors and simulators.

Although little or none of the original code survives today, many of the key ideas survived. Probably the most important idea is that it is possible to build complex chips without armies of workers. In fact, it is common for one or two people to develop a non-trivial, full-custom chip in a few weeks, from ideas to mask, for use in their research projects. For many projects the effort involved in building a custom chip with IDA is comparable to that of building a single TTL breadboard, with the considerable advantage that multiple copies are available with no additional effort or delay.

To make this possible, the CAD environment has grown into a range of tools supporting almost all aspects of design from entry to fabrication, including layout-rule checkers, simulators, routers, *etc.*, all of which are designed to work with each

1

other and present a uniform interface to the designer. These tools embody a set of design principles that have proven to be effective for VLSI CAD. Specifically, the IDA designer relies on:

- Layout by geometric constraint: Designers specify the relative positions of components on the chip symbolically, and the tools determine their final numeric positions.

- Symbolic connectivity: When a design is specified by geometric constraints it is simultaneously constrained electrically. This means that the intended connectivity can be compared with the actual topology resulting in a higher degree of confidence in design correctness.

- Delayed technology binding: Leaf cells are compacted and assembled into larger symbols which represent subsystems on a chip. This compaction step serves several purposes. In particular, it frees the designer from the details of geometric design rules, and makes it easier to parameterize cells both geometrically and electrically.

- Automatic layout synthesis: To get maximum productivity from the limited number of skilled designers available, IDA incorporates layout synthesizers. The early synthesizers were called *generators*. Generators produce regular, or fixed-floorplan cells, such as counters. More recent tools accept arbitrary logic specifications. Practical chips require both.

This hardware design method is backed up by software methods that make the tools easier to develop and use. The key points here are:

- A common language. The IDA tools communicate in the IMAGES design language. Tools incorporate the IMAGES translator to read IMAGES files, build local data structures from them, and write IMAGES files. The semantics of IMAGES make many CAD programs simpler because the programs are provided with a wide range of common functionality without duplicating code.

- A common data structure. All the principal IDA tools, including the graphical editor, the net extractor and the design-rule checker, work from a common internal data format that is built by the IMAGES translator. This commonality has greatly simplified interface building.

- Technology independence. The fabrication technology description is read from a technology file at the start of execution of each program. Programs access technology data exclusively through data structures built from this file. Because the IDA software programs do not "hard-code" technology assumptions, they can be easily ported to new technologies.

- The UNIX$^{TM}$ system. All of the IDA tools work under UNIX and take advantage of its capabilities. This influences both the chip design and the tools themselves. For example, not only are the tools recompiled with the "make" facility,* but most chips are assembled by a series of operations controlled by "make." In fact, several of the tools automatically construct directories of files and a "make" control file to direct subsequent processing steps.

These hardware design and tool-building strategies have evolved over time in light of experience using IDA to build chips.

## Overview of this Volume

This book describes the internals of IDA in some detail. It is intended for the professional CAD worker or student interested in the problems and practical solutions involved in building a real, on-line CAD system. The first thing that the user and CAD worker needs to learn about IDA is the underlying language, IMAGES, which is the subject of the next chapter. A CAD worker may also be interested in how IMAGES is represented internally, and how the constraints are resolved. These issues are also discussed here. Following this is a section on the major IMAGES manipulation tools, including the graphics editor, and geometric manipulators. This is followed by IMAGES synthesis tools, most especially the silicon converter system (SC2D) which is the main layout synthesis system used by IDA today. This is followed by higher-level, circuit-level synthesis discussion. The final chapter is a summary accompanied by a whirlwind discussion of CAD issues and opinions about the future of VLSI.

---

* "make" is a UNIX utility that controls system software based on the modification times of files. It is normally used to automatically recompile programs when source files are newer than the corresponding executable images.

# Chapter 2
# The IMAGES Language

IMAGES serves as the primary medium for *designing* and *describing* the geometric and electrical properties of integrated circuits in the IDA environment. To some extent, the first function, *design*, competes with the second function, *description*. This is because the needs of designers are many and complex: They are involved in the synthesis and the exploration of an unlimited design space. On the other hand, if the only requirement were to describe an existing chip, the language could be simpler, the files smaller and the support software faster. Although one could consider using two different languages for these two purposes, this has been rejected because there is no clear point to draw the line. Any practical chip contains pieces that are old and fixed, while other portions are undergoing frequent human and machine modifications. Because of this dichotomy of purpose, the IMAGES language represents a compromise between ease of use and efficiency.

IMAGES is a human readable, textual language, not a binary format as used in some other systems. This is essential to provide support for human generated IMAGES. As will be demonstrated, even in this age of "silicon compilers" there are some operations that are best controlled by humans working directly in the appropriate higher level language. In addition, some of the silicon compiler systems themselves can be simplified and generalized using IMAGES as a higher-level output language.

Another example of the breadth of IMAGES application is that it supports schematic, virtual-grid and fixed-grid representations of a design. In the virtual-grid mode the user works on a coarse grid where each grid point exactly fits one wire, one contact or one connection to a transistor. This speeds up and simplifies both human editing and automatic cell generation tools since everything automatically fits correctly. Because the user does not have to worry about the spacing of circuit elements or design rules, writing generators in IMAGES is easier than writing generators in "L" [Math83], "HILL" [Leng83], ALLENDE [Mont85] or other languages that do not operate in a virtual-grid design

5

environment.

## 2.1 Background

IMAGES is a successor to the "i" language developed by Steve Johnson [John82], which is discussed in Appendix B. IMAGES' features reflect needs and interests of integrated circuit design groups as well as experience with the Gate Matrix [Lope80], GRED, IDA [Hill84-2] (early) and MULGA [West81] design environments. As a language for writing macro-cell generators IMAGES can be compared to other descendants of "i" including HILL and "L" (from Silicon Design Labs), as well as ALI [LN82] and its successors [LV83] [Mont85].

## 2.2 Principal Constructs

IMAGES programs consist of a list of symbols (which are sometimes referred to in the literature as "cells") which are marked by the keyword SYM. Symbols contain primitive circuit elements including devices such as n and p type MOS transistors marked with the keyword DEVICE, contacts or vias (CONTACT), wires (WIRE), ports (PORT that are sometimes called "pins" or "terminals" in the literature). IMAGES also supports primitive pieces of mask geometry such as polygons (BLOB) and rectangles (RECT). In order to support hierarchy, symbols may also contain instances of previously defined symbols (INST).

Other IMAGES statements exist for manipulating the geometric placement and electrical connectivity of primitive circuit elements. The position constraining statement, BIND, is used to constrain geometric placement of circuit elements. Another statement, PASTE, is used to geometrically constrain and electrically connect instances of symbols. To help specify geometric relations, arbitrary points in the layout may be named with the keyword MARK. A simple example of an IMAGES program is given in Figure 2-1, and the corresponding layout is shown in Figure 2-2a.

```
SYM inv_v IBEGIN
    DEF_NET out_net;
    DEVICE TP top WIDTH=2 ORIENT=VER ;
    DEVICE TN btm WIDTH=1 ORIENT=VER ;
    PORT  POLY in ;
    WIRE POLY WIDTH=1.2 btm.gt1 UP 4 TO top.gt2;
    WIRE POLY in RIGHT 8 TO btm.gt1;
    CONTACT MDP dpout top.drn ;
    CONTACT MDP srcpwr top.src ;
    CONTACT MDN dnout btm.drn ;
    CONTACT MDN srcgnd btm.src ;
    WIRE METAL dpout TO dnout;

    PORT METAL vddleft ;
    MARK METAL vddcenter ;
    PORT METAL vddright ;
    WIRE METAL WIDTH=2 vddleft RIGHT 4
      TO vddcenter RIGHT  8 TO vddright;
    WIRE METAL WIDTH=2 srcpwr UP 8 TO vddcenter;

    PORT METAL gndleft;
    MARK METAL gndcenter;
    PORT METAL gndright;
    WIRE METAL WIDTH=2 gndleft RIGHT 4
      TO gndcenter RIGHT 8 TO gndright;
    WIRE METAL WIDTH=2 srcgnd DOWN 8 TO gndcenter;

    CONTACT MNTUB tubtop vddcenter;
    CONTACT MPTUB tubbtm gndcenter;
    PORT METAL out (dnout,in);
    CONNECT out_net out dnout;
IEND
```

Figure 2-1: An IMAGES Leaf Cell.

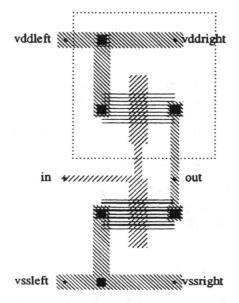

Figure 2-2a: The Corresponding Layout.

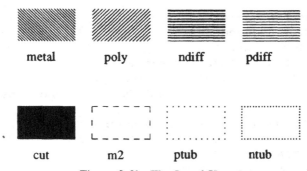

Figure 2-2b: The Level Key.

More detail on language features will be given in the sections that follow.

## 2.3 Geometric Constraints

Fixed coordinates are rarely used in hand-edited IMAGES programs. Tools that generate IMAGES vary in their use of constraints. Some tools, such as the editor, write a dialect of IMAGES that only constrains objects which happen to be geometrically coincident. Other tools, such as the routers, depend on IMAGES to complete their layout work: They generate an IMAGES file that makes extensive use of the geometric and electrical constraint features.

The inverter given in Figure 2-1 is an example of constraint based design: It contains no fixed coordinates at all. Instead objects are placed relative to each other by using the WIRE and BIND statements in the IMAGES language. Geometric constraints in IMAGES are of two types: *equality* and *inequality*. For instance the statement:

```
WIRE METAL dpout DOWN TO dnout;
```

constrains, by virtue of the DOWN keyword, the $x$ coordinates of dpout and dnout to be equal. An inequality constraint is introduced between the $y$ coordinate of dpout and the $y$ coordinate of dnout. The $y$ coordinate of dpout is constrained to be to be greater than or equal to the $y$ coordinate of dnout.

Geometric constraints between objects may be introduced in several ways. For instance, the same constraint-based geometric placement of these elements could be accomplished without adding a wire by the following statement:

```
BIND dpout ABOVE dnout;
```

Alternately, dpout could be constrained to be a distance of exactly three units above dnout by either of the following statements:

```
!equality constraints
WIRE METAL dpout DOWN  3 TO dnout;
BIND dpout ABOVE dnout BY 3;
```

or, at the time of defining dpout:

```
!place the x coordinate of dpout
!at the x coordinate of dnout
!and place the y coordinate of dpout
!at the y coordinate of dnout + 3
CONTACT MP dpout (dnout , dnout + 3);
```

This technique of "design by constraint" is useful for people writing directly in IMAGES since it eliminates the tedious job of recalculating absolute positions whenever there is a minor layout change. It also simplifies the design process when there are no changes. As the resolution of geometric constraints is one of the principal features of the IMAGES language we shall go into some detail on the procedure by which these constraints are resolved to create a layout.

## 2.4  Geometric Constraint Resolution

As mentioned above user-defined constraints in IMAGES are of two types: equality constraints and inequality constraints. The user-defined constraints of an IMAGES program are similar to the design rule constraints that a constraint-based compacter must solve, but several important differences exist. Typically a compacter, working from a "sticks" or virtual-grid design, has an initial layout and can easily find the topmost or leftmost elements of the layout. Locating these elements is useful for ordering  vertices of the design-rule constraint graph associated with the circuit. Since circuit designs in IMAGES have no initial placement, the problem of finding the leftmost or topmost elements of the circuit is equivalent to the problem of finding a feasible layout, which is precisely the problem that the IMAGES constraint resolver is trying to solve. On the other hand, if the initial layout fed to the compacter has no spacing violations (as is the case with a virtual-grid layout), illegal constraints in the design-rule constraint graph should not be present. In a user-defined IMAGES program, however, any illegal constraints may have been mistakenly included in the source file. Moreover, one of the important jobs of the IMAGES constraint resolver is providing intelligent error messages in this situation. A third difference between the problem facing the IMAGES translator and a compacter has to do with the number of fixed objects. In a one-dimensional compaction algorithm the compacter assumes that leftmost and topmost circuit elements are fixed, and compaction is done with respect to those. In comparison, in an IMAGES circuit design, geometrically fixed elements may appear throughout the design, or alternately, there may be no fixed elements in the design at all. Because of these differences, the IMAGES constraint resolver faces a more complex problem than the constraint resolver of an ordinary one dimensional constraint-based compacter. Having motivated the use of geometric constraints we shall describe how geometric constraint resolution is done in the IMAGES compiler.

The most important observation about constraint resolution is that it is equivalent to finding the longest path in a graph. In the remainder of this chapter we will assume a familiarity with basic concepts of graph algorithms. A good introduction to graph algorithms is given in [AHU74]. The intuition behind the relationship between constraint resolution and the longest path problem can be seen in a simple example. Given a set of inequality constraints of the form:

$$v_n \geq v_{n-1} + c_{n-1}$$
$$v_{n-1} \geq v_{n-2} + c_n$$
$$\cdots$$
$$v_2 \geq v_1 + c_1$$

Example 1: a System of Constraints

If we associate the variables $v_i$ with vertices in a graph and the constants $c_i$ with weights on edges between them, then a value of $v_n$ that satisfies the constraints is equal to the value of $v_1$ plus the length of the longest path from $v_1$ to $v_n$. The longest path problem is directly related to the shortest path problem. These observations mean that the rich literature on shortest paths is applicable to the problem of geometric constraint resolution. This observation was made by Lengauer and the paper [Leng84] does a good job of putting the problem in context. We will introduce some further enhancements by Johnson [John77] and which were applied to geometric constraint resolution by Liao and Wong [LiWo83]. The union of these approaches will provide for an efficient solution to the problem.

As there are many good theoretical treatments of shortest path problems [Tarj83], [Mell84] and their application to constraint resolution [Leng84], we will concentrate on the practical issue of trying to find a synthesis of known techniques that provides an efficient solution to the problem at hand.

Example 1 also shows an important difference between constraint resolution and the shortest/longest path problem. In the shortest path problem we typically assume that the objects, such as cities, are fixed and we wish to compute the length of a path. In constraint resolution we are not trying to compute a property of objects with a fixed placement. Instead we are trying to find a placement of objects given constraints between them. The import of this difference is that we must be careful about how we "get started" in placing objects and we must ensure that we place our objects in as small an area as possible. A little care will be given to these issues in the presentation of each algorithm and in the final section we devote considerable effort to resolving all the difficulties that arise from this problem.

## General Environment of Constraint Resolution

The IMAGES statements that produce constraints are processed in one pass by the IMAGES translator. From these statements constraint graphs for the $x$-dimension and the $y$-dimension constraints are built and solved independently. Constraints are made relative to the symbolic coordinates associated with objects, and not with the objects themselves. Thus in the following discussion we need not concern ourselves with transistors, contacts or even with $x$ and $y$ coordinates, but only with an abstract constraint graph $G = (V,E)$. Elements $v_i$ of $V$ are associated with symbolic objects whose value, $\text{VAL}(v_i)$, is an $x$ or $y$ coordinate. Elements $<v_i,v_j,c_{ij}>$ of $E$ represent constraints between these symbolic objects.

## The Solution of Sets of Equalities

The algorithm that we will ultimately be using for solving geometric constraints is built up from a number of algorithms for solving simpler sub-problems. We shall begin with the simplest problem: solving sets of equality constraints. Suppose that we are given a system of equalities in the following form:

$$v_i = v_j + c_{ij}$$
$$v_k = c_k$$

such that each $c_{ij}$ is an integer constant. We wish to assign integer values to each $v_i$ in such a way that all the equalities are satisfied, if such an assignment is possible and to report failure if it is not.

This problem may be solved efficiently by using a union-find algorithm as presented by Tarjan [Tarj75]. A good general discussion of the union-find problem and Tarjan's algorithm in particular, is given in [AHU76].

To see how the union-find algorithm can be applied to this problem, imagine if all the equalities were of the form:

$$v_i = v_j$$

In this case we would simply be creating equivalence classes of coordinates. To generalize this idea to equality constraints with offsets we need only add a little extra accounting information.

We use a tree representation of a set and treat each variable $v_i$ as belonging to a set. Two elements belong to a set if they are related, directly or indirectly, by an equality constraint. Associated with each $v_i$ is a pointer to its parent PARENT($v_i$), a numeric value VAL($v_i$), a weight WEIGHT($v_i$) that gives the number of nodes in the tree rooted at $v_i$, and a displacement value from its parent, DISPLACEMENT($v_i$). Initially each set is a disjoint singleton: PARENT($v_i$) = NULL, VAL($v_i$) = $\infty$ and DISPLACEMENT($v_i$) = $\infty$ .

Getting started with equality constraints is easy. We will assume that every vertex is related to some fixed coordinate or fixed vertex by an equality constraint. If this proves not to be the case we will report an error when "cleaning up" in Algorithm 2-1d. Because we are only dealing with equality constraints, a solution is unique and therefore requires minimal area.

Equalities of the form $v_i = c_k$ are handled by setting VAL($v_i$) = $c_k$. Then in a way analogous to the find portion of the union-find algorithm, this value is propagated up the path to the root of the tree. Key to this procedure is the process of finding the root of a vertex $v_i$. The path from $v_i$ to its root is compressed to improve efficiency. The procedure, "find_root," is given in Algorithm 2-1a. The procedure for making the assignment is given in Algorithm 2-1b. An example of Algorithm 2-1a and 2-1b applied to a constraint graph is given in Figure 2-3.

Equalities of the form $v_i = v_j + c_k$ are handled by taking the union of the sets associated with each of $v_i$ and $v_j$. This is accomplished by translating the equality relation between $v_i$ and $v_j$ into a relation between the root of $v_i$'s tree $r_i$ and the root of $v_j$'s tree $r_j$. To improve the running time of the algorithm, the root of the smaller tree is merged into the larger tree. This process is given in Algorithm 2-1c and an example is given in Figure 2-4.

A final enhancement, unique to this problem, is used. After processing all the equalities we would like the root of each tree to have a fixed (non-infinite) value, if any member of the set (member of the tree) has a fixed value. Fixed constraints given in the IMAGES program must ultimately be honored and we will save processing time if we "bubble up" any hard constraints as soon as possible.

When all constraints have been processed the only thing that remains is to arrive at the final value of all vertices and to check for errors. This process is given in Algorithm 2-1d.

```
/* find the root of v, and collapse the path from v to the root,
** this utility function will be useful both the find and union algorithms*/
find_root(v_i)
    temp = v
    displacement_from_root = 0
    /* find the root*/
    while PARENT(temp) ≠ NULL
        displacement_from_root = displacement_from_root + DISPLACEMENT(temp)
        temp = PARENT(temp)
    end while
    root = temp

    temp2 = v
    /*now collapse path to the root updating the DISPLACEMENT's accordingly*/
    temp_displacement = 0
    while temp2 ≠ root
        temp1 = PARENT(temp2)
        temp_displacement = temp_displacement + DISPLACEMENT(temp2)
        PARENT(temp2) = root
        /* displacement of this node, temp1,  from root is
        ** displacement from v to root - "displacement_from_root"
        ** minus the displacement from v to temp1 - "temp_displacement"  or */
        DISPLACEMENT(temp2) = displacement_from_root - temp_displacement
        temp2 = temp1
    end while
end find_root;
```

Algorithm 2-1a: Find the Root of $v$, and Collapse.

```
/* set v = c and propagate value to root */
find(v, c)
    root = find_root(v)
    VAL(v) = c
    VAL(root) = c - DISPLACEMENT(v)
end find
```

Algorithm 2-1b: Propagating Value to Root.

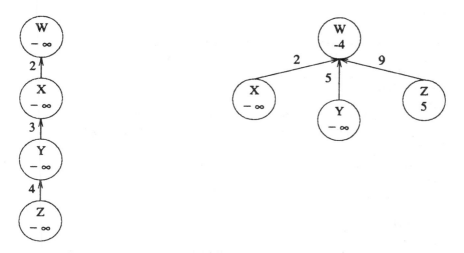

Figure 2-3: Showing the Results of Algorithms 2-1a and 2-1b
Processing Constraint $Z = 5$

```
union(v_i, v_j, c_ij)
/* set v_i = v_j + c_ij */
    r_i = find_root(v_i)   /*VAL(v_i) = VAL(r_i) + DISPLACEMENT(v_i) */
    r_j = find_root(v_j)   /*VAL(v_j) = VAL(r_j) + DISPLACEMENT(v_j) */
    if WEIGHT(r_i) > WEIGHT(r_j) then
        PARENT(r_j) = r_i
        /* VAL(r_j) =
          VAL(r_i) − DISPLACEMENT(v_i) + DISPLACEMENT(v_j) − k  so */
        DISPLACEMENT(r_j) = DISPLACEMENT(v_j) − DISPLACEMENT(v_i) − k
        if FIXED(r_j) then
            VAL(r_i) = VAL(r_j) − DISPLACEMENT(v_j)
        end if
        WEIGHT(r_i) = WEIGHT(r_i) + WEIGHT(r_j) + 1
    else
        PARENT(r_i) = r_j
        /* VAL(r_i) =
          VAL(r_j) − DISPLACEMENT(v_i) − DISPLACEMENT(v_j)+ k  so */
        DISPLACEMENT(r_i) = DISPLACEMENT(v_i) − DISPLACEMENT(v_j) + k
        if FIXED(r_i) then
            VAL(r_j) = VAL(r_i) − DISPLACEMENT(v_i)
        end if
        WEIGHT(r_j) = WEIGHT(r_j) + WEIGHT(r_i) + 1
    end if
end union
```

Algorithm 2-1c: Union.

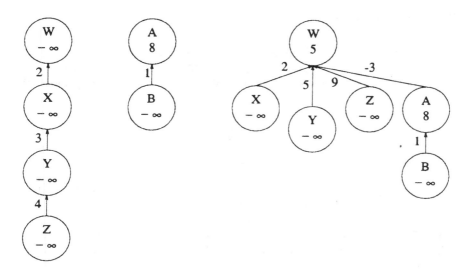

Figure 2-4: Showing the Results of Algorithm 2-1c
After Processing Constraint $Z = B + 5$

```
Clean_up()
    for each v in V
        if VAL(v) = ∞
            root = find_root(v)
            if VAL(root) = ∞
                print "ERROR - unconstrained objects root and v"
            else
                VAL(v) = VAL(root) + DISPLACEMENT(v)
            endif
        end if
    end for
end Clean_up
```

Algorithm 2-1d: Clean_up.

## Solving Sets of Inequalities with Positive Offsets

Solving equality constraints is certainly necessary in the IMAGES translator but inequality constraints are often generated by IMAGES programs as well. We begin our discussion of solving sets of inequalities by considering inequalities of the form:

$$v_i \geq v_j + c_{ij}$$
$$v_k \geq c_k$$

where each of $c_{ij}$ and $c_k$ is a positive integer. To solve these sets of inequalities we will create a graph $G = (V,E)$ where $V = v_i$ and $E$ is defined below. Associated with each $v_i$ in $V$ we have VAL($v_i$), FORWARD_EDGES($v_i$) and INDEGREE($v_i$). Initially VAL($v_i$) = $-\infty$, FORWARD_EDGES($v_i$) = NULL and INDEGREE($v_i$)= 0. Next, for each inequality of the form $v_i \geq c_{ij}$ we set VAL($v_i$) = max(VAL($v_i$), $c_{ij}$ ).

For each inequality of the form $v_j \geq v_i + c_{ij}$ we create an edge from $v_i$ to $v_j$ with weight $c_{ij}$ and increment INDEGREE($v_j$).

Having reflected all the inequalities in the graph we assign values to the vertices that satisfies all the constraints. We begin by finding all the sources $S$ of the graph $G$. These are the vertices $v_i$ in $V$ such that INDEGREE($v_i$) = 0. If none exist then we have a cycle in the graph. A cycle in the graph implies that we have some sequence of constraints:

$$v_{i1} > v_{i2} > v_{i3} \cdots > v_{i1}$$

This set of inequalities has no solution.

Having found the set of sources $S$ we shall require each vertex $v_i$ in $S$ to be contstrained by at least one constraint of the form $v_i > c_j$ . The rationale for this requirement is simply that the extremum of the graph be fixed. This requirement is natural to applications such as constraint based compaction [Wolf86]. In later algorithms this requirement will be relaxed. To solve these inequalities we search the graph in nearly topological order. We find it convenient not to follow topological order strictly, but instead to require that no outgoing edges from a vertex $v$ are traversed before all its incoming edges are traversed. Finally, we must ensure proper behavior in the presence of cycles. Even if there is a source of the graph it may still have cycles. Thus before forward propagating the value of any vertex we check to see if all of its incoming edges have been already traversed once. If they have, then the INDEGREE of this vertex will have dropped below 0.

A minimal area solution is arrived at because we increase the value of a coordinate only to the minimal amount required by the constraints. We give the algorithm in Algorithm 2-2a. Figure 2-5 shows a constraint graph as solved by Algorithm 2-2a.

```
/* Given a directed acyclic graph G = (V,E)
** assign values to each VAL(vᵢ) in V in such a way that each constraint
** vⱼ ≥ vᵢ + cᵢⱼ reflected by <vᵢ,vⱼ,cᵢⱼ> in E
** is satisfied.
** If it is not possible to satisfy constraints report so.
**/
breadth_first()
    Find the sources S of the directed acyclic graph G
        if none exist report "ERROR CYCLE IN GRAPH"
    for each s in S
        forward_propagate(s)
    end for
    for each s in S
        VAL(s) = 0
    end for
end breadth_first
```

Algorithm 2-2a: Breadth-First Search.

```
forward_propagate(vᵢ)
    for each adjacent vertex vⱼ such that <vᵢ,vⱼ,cᵢⱼ> is in E
        INDEGREE(vⱼ ) = INDEGREE( vⱼ ) − 1
        if VAL(vⱼ) < VAL(v sub i ) + cᵢⱼ then
            VAL(vⱼ) = VAL(vᵢ) + cᵢⱼ
            if INDEGREE(vⱼ) < 0 then
                report "ERROR CYCLE IN GRAPH"
                exit
            end if
            if INDEGREE(vⱼ) = 0
                forward_propagate(vⱼ)
            end if
        end if
    end for
end forward_propagate
```

Algorithm 2-2b: Forward Propagate.

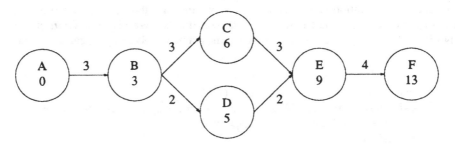

Figure 2-5: A Constraint Graph as Solved by Algorithm 2-2a

Observing that each vertex is traversed at least once and at most $E$ times and each edge is traversed at most twice, the running time of the algorithm is $O(|V|+|E|)$. That is, the running time is linear in the size of the graph and is therefore optimal.

## Solving Sets of Inequalities with Arbitrary Integer Offsets

In the discussion of solving sets of inequalities above we assumed that each edge weight $c_{ij}$ was a positive integer. In many IMAGES programs this will not be the case. In the following discussion we will only require that $c_{ij}$ be an integer. The problem that arises with this generalization is that we may now have legitimate cycles in our graph. When we limited edge weights to positive integers we could be sure that a cycle in the constraint graph implied that our constraints were unsolvable. However, if we allow non-positive edge weights this will not be the case. For example the inequality constraints:

$$v_i \geq v_j + 5$$
$$v_i \geq v_j - 5$$

have a solution, but produce a cycle in our graph from $v_i$ to $v_j$. These cycles mean that it will not be possible to order our graph topologically. If our graph is not topologically ordered then it will be hard to find a search strategy that avoids redundantly visiting edges and vertices. As a result the running time of our algorithm will not be good. A solution that uses no ordering procedure on the graph was independently discovered by Moore [Moor57] and Bellman [Bell58]. A clear presentation of this algorithm is discussed in [Mell84] and [Tarj83]. This algorithm is the simplest to implement and the most generally useful. The idea is that we will initially put each vertex in the queue. At each step of the algorithm we remove a vertex $v_i$ from the queue and visit each adjacent vertex $v_j$. Every time we increase the value of a vertex $v_j$ we enter it into the queue. We will iterate over this procedure until: 1) no further work is done or 2) we have discovered that the constraints associated with the graph are unsolvable. We will need additional fields associated with each $v_i$: COUNT($v_i$) which is used to count the number of times that a vertex enters the queue and IN_QUEUE($v_i$) which is used to determine if the vertex $v_i$ is already in the queue called "QUEUE." Initially QUEUE is empty. During the building of the graph, for each constraint of the form $v_i \geq c$ we set VAL($v_i$) = max(VAL($v_i$), $c$). We shall require that for each path in $G$, there must exist at least one vertex constrained by a constraint of the form $v_i > c_j$ . Then to get started we enter each such $v_i$ in the queue. That a vertex need only be deleted from the queue $|V|$ times is shown in [Mell84]. Thus the algorithm runs in time $O(|V||E|)$.

```
/* for each constraint of the form vi > cj in E*/
solve()
    for each <vi,0,cij> in E
        VAL(vi) = max( VAL(vi), c).
        insert vi in QUEUE
        IN_QUEUE(vi) = true
            end for
    while QUEUE not empty
        remove vi  from QUEUE
        IN_QUEUE(vi)  = false
        COUNT(vi) = COUNT(vi) + 1
        if COUNT(vi) ≥ |V| + 1 then
            report "ERROR - CYCLE in CONSTRAINT GRAPH"
            exit
        endif
        for each <vi,vj,cij> in E
            if VAL(vj) < VAL(vi) + cij then
                VAL(vj) = VAL(vi) + cij
                insert vj in the rear of QUEUE
                IN_QUEUE(v) = false
            endif
        end for
    end while
end solve
```

Algorithm 2-3: General Inequality Constraint Solver.

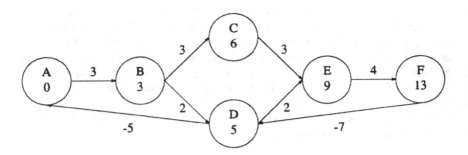

Figure 2-6: A More General Constraint Graph as Solved by Algorithms 2-3 and 2-4

Figure 2-6 gives a more general constraint graph that can be handled by Algorithm 2-3. While Algorithm 2-3 is easy to understand and implement, it is not efficient enough to handle large numbers of inequalities because we use no ordering on the graph whatsoever. We discuss alternatives to this approach in the next section.

**More Efficient General Inequality Solver**

The question arises whether there is an ordering on the graphs $G$ which will allow the graph to be more efficiently searched, at least in an average case. Liao and Wong [LiWo83] propose partitioning the graph $G$ into a graph $G_p$ and $G_n$ where an edge $<v_i, v_j, c_{ij}> \in E_p$ if $c_{ij} > 0$ and is a member of $E_n$ otherwise. The graph $G_p$ can then be topologically ordered and then solved by the breadth-first search algorithm given above. The question then remains how is the influence of the non-positive edges felt in the solution. The approach used is to solve $G_p$, add in the edges of $E_n$ then solve $G_p$ again. This process is applied iteratively until an iteration occurs where no work is done, $i.e.$, when the value of no vertex in $V$ changes. Algorithm 2-4 gives a more complete description of this procedure. We shall require each vertex $v_i$ in $G_p$ with INDEGREE$(v_i) = 0$ to be constrained by at least one constraint of the form $v_i > c_j$. As before, the rationale for this requirement is simply that the extremum of the graph be fixed. In later algorithms this requirement will be relaxed.

```
efficient_solve()
   count = 0
   repeat
     work_done = false
     count = count + 1
     if count > |E_n| + 1 then
        print "inconsistent constraints"
        exit
     end if
     solve G_p using Breadth-First Search (Algorithm 2-2a)
     for each edge <v_i, v_j, c_ij> in E_n
        if VAL(v_i) < VAL(v_j) + c_ij then
           VAL(v_i) = VAL(v_j) + c_ij
           work_done = true
        end if
     end for
   until !(work_done)
end efficient_solve
```

Algorithm 2-4: More Efficient General Inequality Solver.

In [LiWo83] this algorithm is shown to terminate with a solution, if a solution is possible, in $|E_n + 1|$ steps. The proof may be briefly explained as follows. First define $T_k$ as the subset of $V$ such that a vertex $v_i$ is in $T_k$ if, among all the longest paths to $v_i$, the one with the smallest number of non-positive edges has exactly $k$ non-positive edges. Assuming that there are $n$ non-positive edges in $G$, $T = \{T_1, T_2, ..., T_n\}$ is a collection of disjoint sets whose union is $V$. Let $L$ be defined as follows:

$$L = \min \{ u : T_i \text{ is empty if } i > u\}$$

Liao and Wong show by induction that at the $j$th iteration through the solution of $G_n$ followed by the addition of $E_n$ the values of vertices in $T_k$ have settled for $0 \leq k \leq j-1$. Thus after $|L+1|$ iterations the graph $G$ has been solved, if it is solvable. No path in $G$ may have more than $|E_n|$ non-positive edges so $|L| \leq |E_n|$. Thus the running time of this algorithm is $O((|V|+|E|)(|E_n|+1))$.

This approach has the advantage that in the presence of few non-positive edges we have considerably improved over the naive $O(|V||E|)$ solution. The disadvantages of this approach are that it involves separate accounting for non-positive edges and it has a large class of systems of inequalities for which it is inefficient. Particularly poorly handled are systems of inequalities that produce graphs that are topologically orderable but have clusters of non-positive edges. It should be noted that their algorithm was particularly aimed at the problem of solving constraint-based compaction in the presence of user-defined constraints; it was not designed to address a more general class of problems.

A more general context for discussing the impact of partitioning algorithms on the shortest path problem is given by D. Johnson in [John77]. Partitioning $G$ into positive and non-positive arcs is also considered there.

## Solving Systems of Equalities and Inequalities

So far we have considered the solution of equalities and inequalities independently. We now consider problems where both are present, that is, constraints of the following forms:

$$v_i \geq v_j + c_{ij}$$
$$v_k \geq c_k$$
$$v_i = v_j + c_{ij}$$

An obvious approach to the problem is to reformulate each equality constraint so that:

$$v_i = v_j + c_{ij}$$

is re-expressed as:

$$v_i \geq v_j + c_{ij}$$
$$v_j \geq v_i - c_{ij}$$

Such an approach is followed in [LiWo83]. A disadvantage of this approach is that it introduces a negative edge in the graph for each equality constraint. As we saw in the previous section, these negative edges have a multiplicative effect on the running time of constraint resolution and for this reason we have handled negative edges differently.

An approach followed by Steve Johnson in the "i" compiler and also by Lengauer [Leng84] is to treat the equalities and inequalities separately. The equality constraints partition the vertices into a set of equivalence classes. One member of each equivalence class will be chosen to act as the designated representative of that class. Then each inequality constraint will be translated into an equivalent constraint involving only the designated representatives. Equalities are handled by the same technique used in the union-find algorithm. The graph for the inequalities is built just as before. After the equalities and inequalities are processed, a pass is made over all vertices moving the edges from the non-designated equivalence class members to the distinguished member of each class. This process is described below.

> for each $v_i$ in $V$
>   $root_i$ = find_root ($v_i$) /* Use Algorithm 2-1a*/
>   replace each edge $<v_i,v_j,c_{ij}>$ in $E$
>     by an edge $<root_i,root_j,d_{ij}>$ in $E'$ where
>     $d_j = DISPLACEMENT(j) - DISPLACEMENT(i) + c_i$
> end for

Algorithm 2-5a: Moving Equivalence Class Members.

At this point we have a new graph $G' = (V',E')$. This transformation is shown in Figure 2-7.

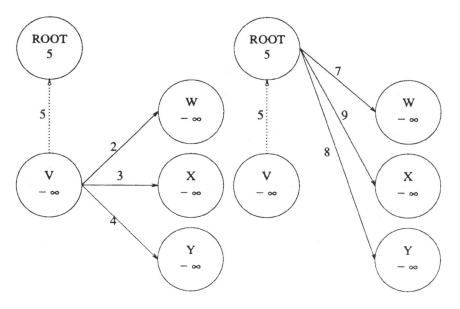

Figure 2-7: Building $G'$ Using Algorithm 2-5a

We then solve this graph using the More Efficient General Inequality Solver given in Algorithm 2-4. Now all the members of $G'$ have their proper value and it only remains to update the values of the members of $V-V'$. This is accomplished by the following.

for each $v_i$ in $V\text{-}V'$
    $VAL(v_i) = DISPLACEMENT[v_i]$
end for

Algorithm 2-5b: Value Update.

Given a Graph $G = (V,E)$, such that $E = E_{eq} \cup E_{ineq}$ expresses a system of equalities and inequalities, the complete algorithm for solving this system is as follows:

Solve $E_{eq}$ using Algorithms 2-1a-d

Re-express each Inequality in $E_{ineq}$ according to Algorithm 2-5a
    and call this $E'^{ineq_{ineq}}$.

Let $V' = \{v_i \mid <v_i, v_j, c_{ij}>$ or $<v_j, v_i, c_{ji}>$ is in $E'_{ineq}\}$.

Solve $G' = V', E'_{ineq}$ using Algorithm 2-4

Solve $V - V'$ using Algorithm 2-5b.

Algorithm 2-5c: Solving Systems of Equalities and Inequalities.

**Solving Fully General Constraints**

In the previous sections we have avoided some classes of constraints and have made other restrictions on the nature of the input. In this section we will provide a final algorithm, using the techniques we have developed so far, that will be capable of handling a fully general set of constraints.

One additional class of constraints that we wish to handle are of the form:

$$v_i \leq c$$

The additional challenge to the constraint resolver given by this class of constraints is that of keeping the placement in as small an area as possible. So far we have accomplished this by starting points at $-\infty$ and moving them no farther in the positive direction than is necessary. An initial placement of $v_i$ at $-\infty$ would satisfy the above constraint, but not in a way that would minimize the area of the placement.

We would also like to be able to handle graphs as general as that given in Figure 2-8. Such a configuration could not be handled by the previous algorithms. The problem with the graph in Figure 2-8 is that only vertex ZERO has an initial value. Using Algorithm 2-4 we would initially solve the subgraph consisting of the positive edges then add in the negative edges. However, because the root of the positive subgraph (vertex Z) is unconstrained, following the topological ordering of the positive subgraph would not result in solving the constraints. Previously we avoided this problem by requiring Z to be constrained, either by an equality constraint such as $Z = 5$ or an inequality constraint such as $Z \geq 5$. Such a requirement is not too restrictive for applications such as compaction where compaction of the layout can be performed relative to fixed left and bottom boundaries. However, this would require that the user to keep in mind fixed coordinates when writing IMAGES programs, and thus would defeat much of the purpose of "designing by constraint."

Fortunately, these problems can be solved by a natural extension of our previous algorithms.

For each constraint of the form:

$$v_j \geq v_i + c$$

we introduce an edge $<v_i, v_j, c_{ij}>$ in $E_{for}$ and also an edge $<v_j, v_i, c_{ij}>$ in $E_{bak}$. This is shown in Algorithm 2-6a. Figure 2-9 shows the backward edges $G_{bak}$ associated with the edges $G_{for}$ of Figure 2-8. Note that the edges in Figure 2-9 are reversed from Figure 2-8 and as a result the fixed vertex, ZERO, is the intitial vertex in the graph.

No special attention need be paid to equality constraints because, as in the previous section, they are factored out of the inequality resolution.

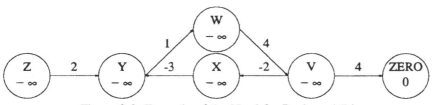

Figure 2-8: Example of the Need for Backward Edges

We now iteratively solve $G_{for}$ and $G_{bak}$. When solving $G_{for}$, using Algorithm 2-4, we maintain the invariant that no edge gets moved farther than necessary in the positive direction. When solving $G_{bak}$, using a modified version of Algorithm 2-4, we maintain the invariant that no edge gets moved farther than necessary in the negative direction. The entire algorithm is given in Algorithm 2-6b. Using Algorithm 2-6b we ensure that the placement requires as little area as possible and that the constraint resolution is performed in an efficient way.

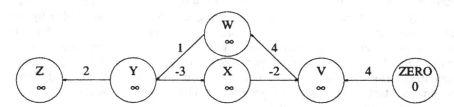

Figure 2-9: $G_{bak}$ Associated with Figure 2-8

for each edge $<v_i, v_j, c_{ij}>$ in $E_{for-ineq}$
    create an edge $<v_j, v_i, c_{ij}>$ in $E_{bak-ineq}$
end for

Algorithm 2-6a: Re-expressing Inequalities.

Solve $E_{eq}$ according to Algorithm 2-1

Re-express each Inequality in $E_{ineq}$ according to Algorithm 2-5a
and call this $E_{for-ineq}$.

Re-express each Inequality in $E_{ineq}$ according to Algorithm 2-6a
and call this $E_{bak-ineq}$.

Let $V' = \{v_i : <v_i, v_j, c_{ij}>$ or $<v_j, v_i, c_{ji}>$ is in $E_{for-ineq}$ $\}$.

Partition the edges in $G'_{for} = (V', E_{for-ineq})$ ,
  into positive and negative edges

Partition the edges in $G'_{bak} = (V', E_{bak-ineq})$
  into positive and negative edges

While *work_done*
  Solve $G'_{for}$ using Algorithm 2-4
  Solve $G'_{bak}$ using Algorithm 2-4 but
    in Algorithm 2-4 and Algorithm 2-2b replace the lines:
    if $VAL(v_i) < VAL(v_j) + c_{ij}$
        $VAL(v_i) = VAL(v_j) + c_{ij}$
    by
      if $VAL(v_i) > VAL(v_j) - c_{ij}$
        $VAL(v_i) = VAL(v_j) - c_{ij}$
end while

Solve $V-V'$ using Algorithm 2-5b.

Algorithm 2-6b: Solving General Systems of Equalities and Inequalities.

## Summary: Efficiency of Constraint Resolution

The approach to geometric constraint resolution described in this section reflects a comprehensive survey of the relevant literature, as well as a mature perspective on the practical impact of various algorithmic alternatives. In our implementation of the IMAGES translator several attempts were made to simplify the final algorithm. For instance we tried eliminating the special handling of equality constraints that marks the transition from Algorithm 2-4 to Algorithm 2-5c. To date, all such simplifications have been rejected due to their negative impact on performance. The practical results of our effort are that constraint resolution has always run faster than parsing or the construction of the basic data structures. While we can claim no significant theoretical advancements on the underlying problems of constraint resolution, we hope that our synthesis of existing techniques will be useful to those seeking to solve other VLSI design problems such as compaction, as well as unrelated problems involving constraint resolution.

## 2.5  Electrical Connectivity

IMAGES provides constructs for specifying the electrical connectivity of a circuit design and uses this information to make inferences about the circuit when processing the design. This facility does not replace the need for circuit extraction but rather complements it. The IMAGES view of electrical connectivity can be checked against the extracted view to ensure that the connectivity reflects the designer's intention.

Inside IMAGES a *net* is a property associated with a set of IMAGES *connections*. Connections represent electrical connection points in the three dimensional design space. Connections are associated with terminals of devices, contacts, ports and marks of symbols. Connections differ from simple geometric locations in that they have a set of compatible layers and a net associated with them.

The IMAGES translator has two rules for deriving electrical connectivity from information supplied by the user:

1.  If an IMAGES object that is a member of a net is placed at a connection, and if the layer(s) associated with the connection are consistent with the layer(s) associated with IMAGES object, then a single net results that is the union of the nets associated with the connection and the object.

2.  If a wire connects two or more connections, and if the layer(s) associated with the connections are consistent with the layer associated with wire, then a single net results that is the union of the nets associated with each of the connections.

As an example of connectivity derivation, consider the following portion of Figure 2-1:

```
DEVICE TP top WIDTH=2 ORIENT=VER ;
PORT METAL vddleft  NET=vdd;
MARK METAL vddcenter ;
PORT METAL vddright ;
!place the contact at the diffusion west
!terminal of transistor top
CONTACT MDP srcpwr top.drn ;
WIRE METAL WIDTH=2 vddleft RIGHT 4
    TO vddcenter RIGHT  8 TO vddright;
WIRE METAL WIDTH=2 srcpwr UP 8 TO vddcenter;
```

After the IMAGES translation of this design `srcpwr`, `vddleft`, `vddcenter`, `vddright`, and `top.drn` will all share the same user-defined electrical net `vdd`. The contact `srcpwr` and the transistor port `top.drn` will share the same net because `srcpwr` was placed at `top.drn` and the first rule comes into play. The connections `srcpwr`, `vddleft`, `vddcenter`, and `vddright` will all share the same net because they are wired together and the second rule is in effect.

As mentioned above, electrical nets may be considered as sets, and connections as set elements. Thus it is natural to use a union-find algorithm [Tarj75] to dynamically resolve the electrical connectivity of a circuit described by an IMAGES program. The approach is precisely that of Algorithm 1. The advantage of using a union-find algorithm over simply finding the connected components (*i.e.* nets) of all the nodes (*i.e.* connections) in the circuit is that errors in the user program may be discovered more quickly and more carefully related to the particular element of the user program that is responsible for the error.

## 2.6  Other IMAGES Features

### Attributes

Each IMAGES primitive circuit element has several *attributes*. The IMAGES language provides constructs that allow the user to query the values of attributes of instantiated cells. For instance, for an instantiated symbol `c` with port `p` the expressions:

```
c.p'X
c.p'LEVEL
```

give the value of the $x$ coordinate of $p$ and its layer respectively. Supported attributes in IMAGES include the geometric coordinates of an object, its layer, net, width and length.

Information about lower-level symbols is particularly useful. Before these attributes were available, a clumsy system of partial compilation and system calls was used to pass geometric information up the symbol instance hierarchy. This was needed to allow higher-level symbols to control their activities depending on the size and shape of the symbols under them. By enhancing the language only slightly, the description became more terse and design process became simpler.

**Loops, Expressions, and Variables**

Since many designs exhibit a high degree of regularity, the IMAGES language allows the use of loops as abbreviations. Although they appear to be similar to loops in ordinary programming languages such as ADA, they are really just short-hand for the same statements written out explicitly. Each time the loop interpreted it is essentially "unrolled." Naturally, to make this work, there must be a loop variable that is incremented as the loop is interpreted and the language must provide references to names based on the value of that variable. In IMAGES, these facilities are all patterned roughly after their ADA counterparts. An example of these is shown below:

```
FOR port_num in 0 .. 4 LOOP
   IF port_num MOD 2 == 0 THEN
      port poly out[port_num] (12,port_num*4);
   ELSE
      port metal in[port_num] (12,port_num*4-4);
   END IF;
END LOOP;
```

These facilities are implemented as an extension of the lexical analyzer. The lexical analyzer accepts tokens from the input file and gathers them together. When a complete context has been seen they are passed to the parser for compilation. In straight IMAGES text without preprocessor dependencies these contexts are just one token long. But when the lexical analyzer detects a loop statement, it gathers together all the tokens from a loop into memory, interprets the loop and produces output proportional to the number of iterations specified. This system complicates lexical analysis with semantic problems. It also has to deal with nested loops and user errors. But it has the advantages that it works quickly in general use, has low overhead when looping features are not used and provides a high degree of flexibility.

## Technology Independence/Technology Accessibility

To make the language as technology independent as possible, IMAGES uses a technology file reader and technology database. The technology file influences the IMAGES translator in two ways: it determines the set of technology words, such as METAL, that will be recognized and it provides some quantitative access to the underlying technology. This complements the use of the compacter: Many global positioning tasks and wiring need be parameterized by only a few simple constants to achieve adequate technology updatability. To accomplish this, a set of technology specific variables and functions are supported such as TECH_MIN_SEP, TECH_LAMBDA, TECH_DEF_WIDTH[METAL] and others. The set of technology variables to be provided is supplied by the technology database, which means that new categories of information can be added with little effort.

## Summary of the Various Uses of IMAGES

The relationships among the ways that IMAGES is typically used is summarized in the following table:

|                  | VIRTUAL                              | FIXED                |
|------------------|--------------------------------------|----------------------|
| CONSTRAINT-BASED | output of generators and routers     | chip assembly        |
| NUMERIC          | translation from outside environment | output of compacter  |

## 2.7 Interface with the Rest of IDA

A typical design path for using the IMAGES language to build a chip might include invoking an *awk*-like or C-like pre-processor whose output would be the IMAGES language. The compacter and optionally the suite of symbolic routers would then be invoked to produce a fixed layout. Once the design is settled, this process may be coordinated using UNIX tools, especially the *make* program. This is discussed in Chapter 5.

## 2.8 Quick Access to Large Designs

In most chip designs, the lower level cells are created graphically and/or a compacter. In either case, the positions and electrical nets associated with their ports are fixed when the symbol is written. The IDA system can take advantage of this property to speed up access with a "pseudo-database" that provides quick, random access to the individual cells. Symbols with fixed ports normally reside in their own files. A design can be spread out over many UNIX directories. This feature is heavily used when libraries of standard symbols are needed or when several people cooperate on a design. The text in these cells is a dialect of IMAGES, formated in a way that allows the external information about a cell to be accessed independently from its internal details. An example of this format is summarized here as it might appear in the file "buffer.im" and "inverter.im":


"buffer.im"


```
        USES inverter;
        SYM buf IBEGIN
                PORT in NET=in (-4,0);
                PORT out NET=out (4,0);
                PORT vdd NET=vdd (4,4);
                PORT vss NET=vss (4,4);
                BOUNDARY UNIVERSAL (0,0) (4,0) (4,4) (0,4);
                ENDHEADER
                INST inverter inv [in=in, out=out] (0,0);
                IEND
```

"inverter.im"

```
USES ;  !the absence of any symbol names here
  !indicates that the ''inverter'' is a leaf cell
SYM inverter IBEGIN
   PORT in NET=in (-4,0);
   PORT out NET=out (4,0);
   PORT vdd NET=vdd (4,4);
   BOUNDARY UNIVERSAL (0,0) (4,0) (4,4) (0,4);
   ENDHEADER;

   TP tp0 (2,3);
   CONTACT MP mp1 (3,3);
   WIRE POLY FROM tp0.gt1 DOWN TO mp1;
   !more wires and contacts, etc.
   IEND;
```

When the symbol "inverter" is referenced in the USES clause of symbol "buf" the IMAGES translator stops reading "buf.im" and reads the file "inverter.im" up to the ENDHEADER line. This provides the ports and overall size of the symbol, which is all that is required to complete the parsing of symbol "buf." If more information is required, the system reinvokes the IMAGES compiler to complete the processing of symbol "inverter." This time it skips over the header and reads the remainder of the symbol definition. Because this format limits the amount of information that must be processed to enter the tool, start-up time is independent of the overall size of the chip.

### Advantages of a Text-Based Design Format

Because the basic IMAGES medium is textual, it is possible to use a wide range of conventional text manipulation tools on it. For example, under the UNIX operating system a command called "grep" is available that searches for a pattern occurring in any specified set of files. To find files that reference the symbol "inverter", the command

```
grep inverter *.im
```

can be used. Other tools include the "awk" language, tools for detecting difference in files and even the mail system and editors. All of these tools are optimized for text files.

The alternative, a binary database, was explored some years ago but it turned out to be unpopular. Several factors led to removing the binary form from the system. The most important was that it was largely redundant. When people design directly in a textual language the "official" design resides in text files and the binary format merely represents a copy of the design. Since the speed of parsing is such that most tools spend less than 10% of their time reading in a design, the extra step of converting to the binary form and additional storage required was never justified. Since the range of algorithmic complexity that is possible in text files (including loops, arithmetic expressions and conditional compilation) far exceeds the capacity of today's binary formats, it is unlikely that human generated files will be stored in binary format soon.

## 2.9  The IMAGES Implementation

The IMAGES translator shares and primarily works from the IDA data structures described in Appendix A. However, some of the key algorithms of the IMAGES translator, such as electrical connectivity extraction and geometrical constraint resolution deal only with specialized geometric and electrical information. To improve their efficiency, a data structure local to the IMAGES translator is created. The principal element of this data structure is called the *connection*. Connections are symbolic coordinates in three-dimensional design space; their third dimension being a set of compatible layers and a net associated with them. Connections are associated with terminals of devices and contact cuts, as well as with ports, marks and symbol instances.

The key point about this data structure is that it provides a single, uniform data structure for performing the principal tasks of the IMAGES translator. Once the connections are built, the constraint resolver need not examine any other data structures. Thus both the electrical connectivity extractor and the geometric constraint resolver may operate in a "sea" of connections and not worry about other objects such as devices, contacts and so forth. This issue is discussed in greater detail in Appendix A.

## 2.10  Summary

The design requirements for IMAGES were many and complex, and to some extent, the final language reflects that complexity. This was essential in order for IMAGES to support its demanding application: the human writing IMAGES directly. Nevertheless, the efficiency of IMAGES has turned out to be adequate to support the other, less demanding, role as a chip description vehicle.

# Chapter 3
# Designer Interaction

Even in a CAD system based on a textual language like IMAGES, graphical support remains important. Many aspects of a design are easier to express, understand and manipulate graphically than textually. The editor called "icon" is the primary tool for graphical interaction with IMAGES designs. It is usually the first tool a designer learns, even if he or she intends to use generators and text-based layout. This is because it provides the most immediate feedback and verification of the other systems. In addition, the editor translates graphically entered layouts into IMAGES, which can provide the user with a starting point for hand editing.

## 3.1 Basic Editing Environment

The structure of the editor emphasizes large-scale and comprehensive manipulations rather than design entry, since most designs are entered once but manipulated many times. Unfortunately, this can create a bit of a hurdle for the first-time user who may be intimidated by the range of options. (The system has over 230 commands.) But chip design is not a quick and casual activity, and most users find that the long-term gain exceeds the short-term learning cost.

Actions in the editor center around the *cursor* and the *mark*.* The Cursor is drawn on the screen as a Cross, the mark as an X (mnemonic: "X marks the spot"). The cursor and mark can be moved using the mouse or the keyboard. For random moves across the screen, the mouse is more convenient. For controlled short moves (such as exactly 8 microns right) or for moves off the currently displayed screen (go to the transistor named "tp3") the keyboard is more convenient. Because the cursor and mark can be controlled by the keyboard, it is

---

* This mark should not be confused with the concept of MARK in the IMAGES language. The naming is a historical coincidence.

possible to do some amount of editing without a graphical screen at all. For example, the user can move the cursor to a named point (such as "reg5.input"), delete everything to the right of it and write out the symbol without ever looking at it. (This method has been used more than once by designers working remotely on a non-graphical terminal.) When the editor is invoked, it puts the cursor and mark in the lower left and upper right quadrants (Figure 3-1).

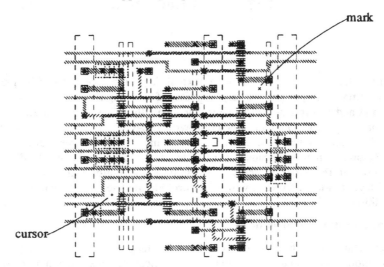

Figure 3-1: Initial Icon Screen Display.

Commands can be entered in three modes, via the keyboard (with mouse for positioning), via a pop-up mouse menu (with some keyboard entry) or from a batch file of keyboard command characters. The first is by far the most popular, the second has some advantages for inexperienced users. The batch file mode is mostly used with prepared scripts to perform modifications on a chip or to test and demonstrate the editor. Overall, the keyboard interaction mode is structured something like the "vi" text editor. Short, relatively innocuous commands, such as 'j' to move the cursor down, are entered in the "raw" mode and cause action immediately. Longer commands, and ones with greater consequences, are initiated with a colon, after which the user can type the command using backspaces to correct mistakes. For example, to write the current symbol in the existing file "register.im" one would type

        :w! register.im

Of course, the analogy with a text editor breaks down quickly, since text editors do not deal with transistors, electrical nets, *etc*. But wherever there is an overlap in function, the text editor command sequence was used.

The editor is structured like a "reverse-Polish" calculator. One first specifies what objects are to be manipulated then specifies the operation. All operations that change the design center around a group of objects called the "chosen." The chosen objects can include wires, inter-level contacts, transistors, symbol instances, internally named points (MARKs) and external definition points (PORTs). The chosen can also include *fragments*, meaning ends of wires, connection points to transistors and ports on symbol instances. Each end of a wire may be manipulated independently, thereby allowing "rubber-banding" of wires. Device and symbol instance connection points are important because the editor keeps track of the electrical net associated with each object in the design. This allows icon to tell the user about the connectivity it understands. At the same time the user can specify that things that appear unrelated are electrically connected. The net information can also be used in editing. For example, the user can have all the features associated with a net highlighted on the screen and placed in the chosen group for further manipulation.

There are many other ways to specify what objects are to be "chosen." Most of the mechanisms involve the cursor. For example, where a transistor, wire and contact overlap, the user may want all three objects or just one. If the user wants just one object it can be selected by category (*e.g.* "identify wire"). If several wires appear under the cursor, the editor selects one arbitrarily, and reissuing the command rotates among them. The user can also identify regions delimited by the cursor and the mark. Objects can be put in the chosen group either inclusive or exclusive of wires passing through the boundaries. Objects can also be selected by naming them, by association with previously selected objects and by other ways that proved handy (*e.g.* the whole design, everything to the right of the cursor, *etc.*). This relationship is shown in Figure 3-2.

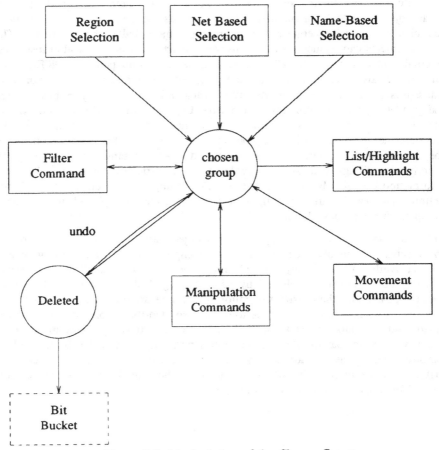

Figure 3-2: Manipulation of the Chosen Group.

Once the chosen group has been selected it is highlighted on the screen. The contents of the group of chosen features can then be adjusted by picking one or more items using a menu or by using the "filter" command. This command allows some of the chosen objects to be culled out of the chosen group by category. For example, the user could select all the objects on the right half of the screen, and then filter out everything but the transistors. Filtering can also be done by polarity, so it is easy to change the dimensions of just the nFETs, for example. One simply selects the entire symbol (with the identify all command) and then filters out everything that is not a transistor. Next the pFETs are filtered out. Finally, a "resize" command will change the current size of the remaining nFETs to the newly specified size. This process is illustrated in Figure 3-3.

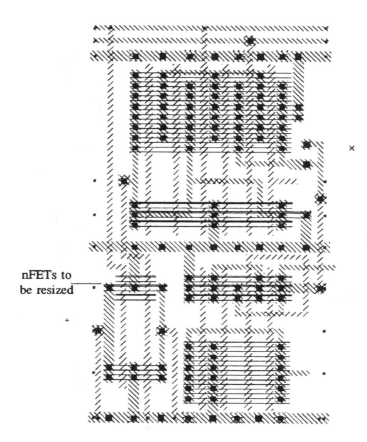

nFETs to
be resized

Figure 3-3a: Before Resize Command.

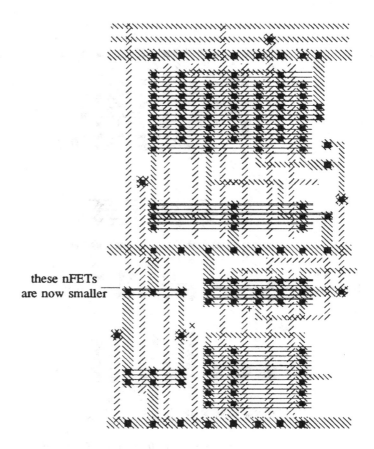

these nFETs
are now smaller

Figure 3-3b: After Resize Command.

This pre-order command structure requires some training, but experienced users can often get exactly the items they want into the chosen group by a combination of stretching items in or out of a region and using inclusive or exclusive enclosing boxes. For example, suppose that the designer has selected a small group of features that form a useful subcircuit such as a buffer. Several copies of the buffer are needed, but the ports associated with the power and ground are not since power connections need only appear once for the whole group of buffers. To build the circuit the user could select the group of objects by indicating a box around them and then filtering out the ports. Now the chosen group can be replicated.

A simple editing process is illustrated in Figure 3-4. Here the designer is trying to create an exclusive-OR gate out of four NAND gates (Figure 3-5) starting with some geometry borrowed from another symbol. (In normal practice borrowing geometry is common, but this is a little extreme since the starting geometry is not closely related to the final result.)

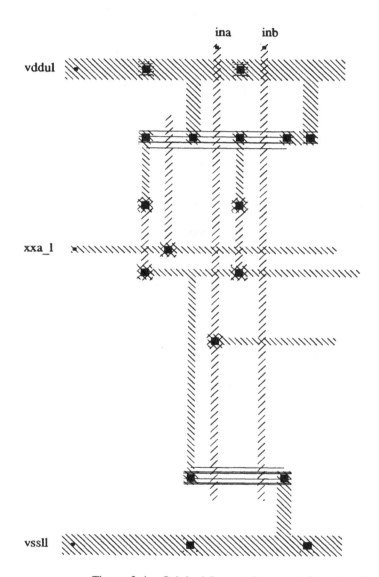

Figure 3-4a: Original Layout Borrowed from Another Design.

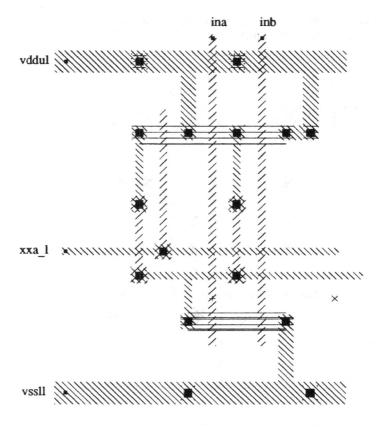

Figure 3-4b: Layout Pruned Down.

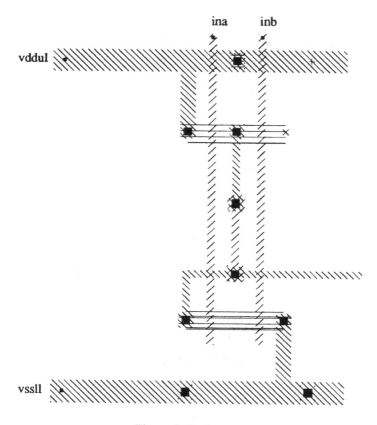

Figure 3-4c: Layout Ready for Replication.

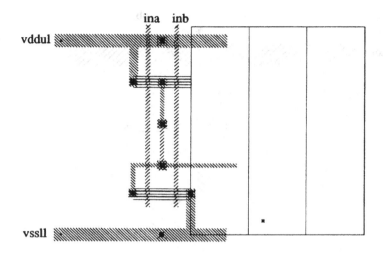

Figure 3-4d: Replicated Layout.

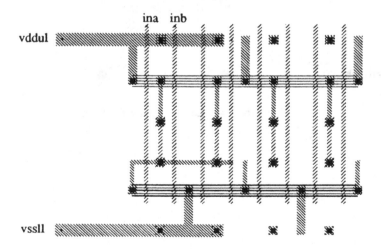

Figure 3-4e: Flattened Layout.

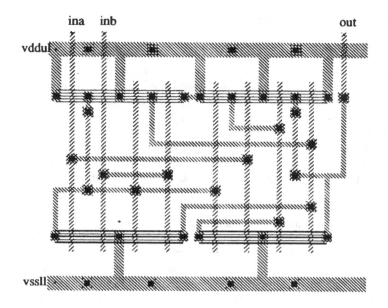

Figure 3-4f: Touched up.

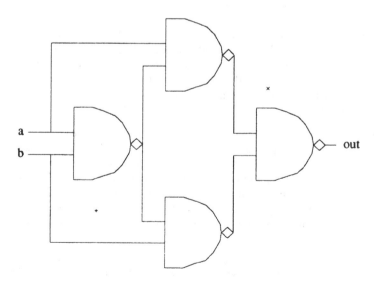

Figure 3-5: Schematic for XOR Gate.

## Commands that Operate on the Chosen

The editor provides many commands that act on all the chosen objects at once. Perhaps the most commonly used is "delete." This command takes all the chosen objects, removes them from the symbol, stores them in an internal group called "deleted" and sets a flag to indicate what type of command was executed (e.g. insert, deletion, movement, *etc*). When an "undo" is requested, the hidden group is selectively added to, removed from or used to un-move the objects in the current symbol. The "deleted" group is preserved in the face of simple commands such as cursor movement or panning the screen. However, if the user deletes something else, the "deleted" group is purged and refilled with the newly deleted objects. It is then impossible to undo the first deletion. This structure seems to be satisfactory for most designers. It allows them to undo most critical operations quickly and avoids unnecessary overhead. (For example, it does not copy the complete design on every command.) Furthermore, the editor understands that many commands (such as replotting commands or commands to change settings) do not influence the chosen group. They cannot be undone, but neither do they lose the context of the chosen group. This means that the user can delete a collection of items, zoom in to inspect the consequences, ask for interactive help, change the settings and still be able to undo the deletion.

Two other heavily used commands that operate on the chosen group include "move" and "copy." Both of these take their direction of movement by the vector from the mark to the cursor. That is, to move a transistor from one spot to another precisely, the following sequence is used: The designer first selects the transistor, then puts the mark on some reference point on the transistor such as the right gate port. He or she then moves the cursor at the point where the reference point is supposed to go such as the left gate port on another transistor. Finally, the designer issues the move command and the transistor moves to the new location. The mark is moved to be coincident with the cursor.

This works just as well with contacts, wires, symbol instances, *etc.* or collections of these objects. This approach should be contrasted with editors that have a "move transistor" command, a "move contact" command and so forth. Icon has only one "move chosen" command*. Furthermore, when an object has been moved (or copied), it remains in the chosen group (or the copy is now in the chosen group) and the mark is put on top of the cursor. This makes it easy to move an object repeatedly which is especially useful if the designer misses the

---

* Actually, it has three. Move by the vector from the mark to cursor (the standard one) and move by the horizontal or vertical component of the vector. The second and third make it easy to line things up accurately.

desired location by a little the first time**. (*I.e.* the designer can say "copy," "move," "move.")

Icon supports many more complex commands that operate on the chosen. For example, the chosen group can be replicated in any direction. Icon will automatically figure out the spacing between copies so that the edges just touch each other. If the chosen contains a series of wires butted end to end, the "fuse" command joins them into one (whenever the wires have the same level, electrical net and width). This proved useful for simplifying an automatically generated layout that consisted of a regular array of cells filled with short segments of wires placed adjacent to one another. Other examples include the "rewidth" and "relevel" commands, which can change the width or level of all the wires in the chosen group. This, combined with the ability to choose all the objects associated with a given net, gives the editor the ability to adjust all power or ground wires simultaneously.

Many of the more powerful commands involve manipulating symbol instances. For example, the user can choose one or more symbol instances using any of several selection commands and merge them into the current symbol. Alternatively, the user can select a group of objects and create a sub-symbol out of them. These two commands can be used to provide the ability to rotate collections of objects as a group. This is accomplished by first chosing the desired group of objects and creating a sub-symbol out of it. Then the user puts one or more rotated/reflected instances of the newly created symbol at the desired location and merges them back into the original design.

Another example of the interaction between the use of the chosen group and the symbol hierarchy is the command "fillports." To understand the need for this command, consider the construction of a 16 bit register out of flip-flops. The first step is obvious, the user designs a flip-flop and puts it in a symbol called, for example, "fflop." The designer then creates a new symbol, for example, "reg16" and places 16 instances of the flip-flop next to each other. But the new symbol must include ports to provide access to the lower level symbol. This could be done textually in the IMAGES language by specifying PORT statements that were symbolically bound to the lower level cells' ports. To do it graphically without specific support would be very tedious: One would have to add a port on top of the input and output instance ports of each cell. But this can be accomplished using the following sequence:

---

** In order to support this, the "undo" command also puts the mark back at its original location.

```
identify all
filter out all but poly instance ports
run the "fillports" command, which
          for each instance port in the chosen group,
          creates a new port in the higher level symbol,
          with the same name if possible, coincident,
          and with the same level
extract electrical nets
```

The result will be 32 new polysilicon ports on top of the corresponding ports in the lower-level symbols.

## Nets and Extraction

Early versions of the editor attempted to maintain electrical information about each object constantly, a feature that is still found in some other systems, especially schematic capture systems. However, in icon this feature was eventually dropped. This was because it proved almost impossible to keep accurate connectivity in the face of powerful commands that can create or destroy thousands of shorts or opens in one keystroke (such as the sequence: "identify all, flatten hierarchy, replicate right 99 copies, undo, undo"). Furthermore, with experience it became clear that designers were not particularly interested in this feature, since many manipulations involve intentionally creating shorts or opens that would later be corrected. Finally, with the advent of the hierarchical, scan-line extractor built into the editor, there was no need to do a "mini-extract" after each command. A complete extraction can be finished in just a few seconds; usually in about the time it takes to draw the symbol.

## Producing IMAGES from Graphics

The extractor algorithm will be discussed in the next chapter, but it is appropriate to mention here how the editor uses electrical information in formatting the IMAGES it produces. When geometry is created or modified graphically, the symbolic dependency information is missing. For example, if the original IMAGES source file specified:

```
DEVICE tp tp1;
CONTACT mdp c1 tp1.drn;
```

When displayed in the editor, one would see that contact called "c1" was directly on top of the drain port of transistor "tp1" because the symbolic dependency forced them to be coincident. But if one were to move the contact to the source port in the editor interactively, this relationship would be lost. The question is: How should the symbol be written out so to provide sensible symbolic relations? One could consider building an "expert system" that would attempt to intuit dependencies among objects based on their proximity (in fact, such as system was once built for IMAGES), but the dependencies are rarely particularly meaningful. The solution that was finally adopted is based on the use of the net information and the inverted net list. The inverted net list is constructed by adding each object to a list of objects on its net. The algorithm (3-1) works as follows:

> extract the electrical net information
> invert each net (attach a list of objects to it)
> For each net use the inverted net list to:
> Print the ports with numerical positions.
> Print the symbol instances with symbolic positions,
>     keyed to the ports if possible, otherwise numerical positions.
> Print the marks with symbolic positions,
>     keyed to the above objects if possible, otherwise numerically.
> Print the devices with symbolic positions,
>     keyed to the above objects if possible, otherwise numerically.
> Print the contacts with symbolic positions,
>     keyed to the above objects if possible, otherwise numerically.
> Print the wires with symbolic positions for each endpoint,
>     keyed to the above objects if possible, otherwise numerically.

Algorithm 3-1: The Procedure Used for Finding Symbolic References.

The algorithm works quickly because it partitions the symbol by electrical net. Most nets, except Vss and Vdd, have at most a few dozen objects on them. The check for coincident positions is fast but is complicated slightly by the need to consider electrically connected levels, such as putting a poly wire symbolically on a metal-poly contact. It produces reasonably compact IMAGES output because whenever it finds a coincident relation it ties the two objects together symbolically. This means that the output file does not need to include a net label on the new object since its net information can be derived by the compiler. When numerical positions are used, a net must be explicitly attached and the output is both more voluminous and less enlighting.

## 3.2  Hybrid Logic/Layout Diagrams

Most VLSI CAD editors support either schematic entry (such as Draw [Fras78]) or layout entry (such as KIC [Kell81]). This dichotomy is appropriate when the schematic is produced by one group of designers and "tossed over the wall" to the layout organization. But when one person is responsible for both aspects, logic and layout must be considered concurrently. Using two tools can lead to conflicts where one tool is tuned to handle schematics and does not understand the properties of layout (such as the various levels of interconnection used) and another is tuned for layout and cannot handle schematics. A further problem with conventional tools occurs when a schematic must include a large block such as a PLA, ROM or RAM for which there is no logic diagram available. Such structures are often best specified in logic equations or state tables and then laid out algorithmically. The conventional solution is to place them in the diagram as a box and to attempt to develop a simulation model by some other means. This can lead to errors when the simulation of these automatically generated blocks does not exactly match their true functionality.

To avoid the overhead of changing from one facility for schematics to another for layout, the editor was designed to support both. The logic designer can use the editor to hook up logic cells from one or more libraries and see them on the screen in the conventional format. In a logic design, the emphasis is on human legibility and understandability. The layout designer can deal with stick diagrams or colored rectangles and sees the layout with accurate dimensions. Here the emphasis is on geometrical constraints and the visual clarity is poorer. In some custom designs there will be a correlation between the schematic and the layout, since in both cases an effort is made to put related circuitry close to each other. But, in general, the process of transforming a schematic into a layout will result in a new relative positioning.

Because the internal structures the editor uses for schematics are the same as those used for layouts, it is possible to intermingle the two. This gets around the problem of interfacing schematics with automatically generated layouts. Once the output of a layout generator is available, they can be combined directly into the logic diagrams. This form of hybrid logic diagram reflects the logic of the final layout and also provides some insight into the floor plan of the final chip. An example of this is the "filter chip" schematic shown in Figure 3-6. This chip consists of a large shift register (in the lower right) combined with an address storage and recognition unit (in the upper left). The shift register was considered so simple schematically that no logic diagram was ever made for it. Rather, a layout was produced semi-automatically by a shift-register layout generator. Then the layout was included into the schematic for the overall chip.

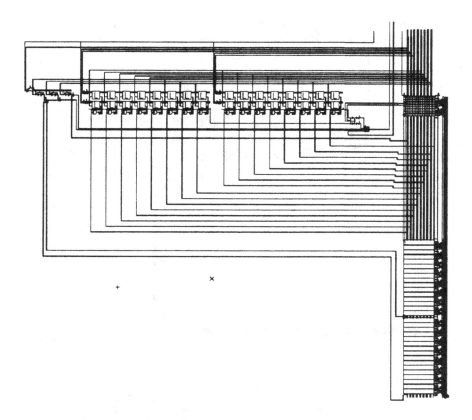

Figure 3-6a: Hybrid Layout-Schematic.

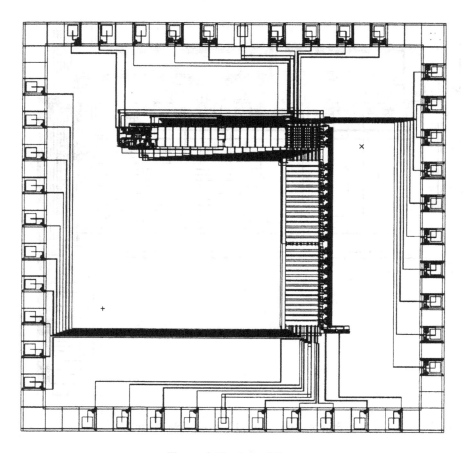

Figure 3-6b: Actual Layout.

The combination of logic and layout in one drawing is supported by several features of the editor. First, each symbol in the design hierarchy may have an arbitrary shape associated with it, preferably one that is reminiscent of its function. These shapes are called "icons" (used in Figure 3-5) and give the editor its name. For low-level symbols these include the traditional NOR, NAND and XOR gates that designers are used to. For somewhat higher-level symbols arbitrary shapes (such as squares with notches cut out or a box with a giant letter "L" in it) form a useful memory aid. For very large symbols, such as a RAM produced by a layout generator, the enclosing box is normally used since the name of the device is the most useful factor.

The second feature supporting logic and layout is a "universal interconnect level" (called UNIV in IMAGES) which forms electrical connections to itself or to any of the physical levels in the design (metal, polysilicon, *etc.*). This is needed because logic designers (as opposed to layout designers) do not know or care what physical level they will use, but may wish to connect to (say) metal and poly lines coming from a RAM. In order to be useful the schematic properties of this universal level must be supported by the IMAGES compiler and the extractor. The extractor (which will be discussed in greater detail in Chapter 4) must allow wires to cross without connections when extracting a schematic. No matter what level the wire are, they are considered connected only when

1.  one of them has an endpoint exactly on top of the other, or

2.  they have a MARK or PORT marking their intersection.

Not only is this required for supporting the schematics mode of the editor, it turns out to be a natural mode to extract virtual-grid layouts as well, because it assists error detection. Although it means that the virtual-grid designer must enter the wires more carefully and must sometimes enter a single long wire as two shorter ones, the payoff is an increased confidence that there are no unintentional connections. In addition, this is more in keeping with the semantics of IMAGES, where wires are connected by symbolic coincidence.

### 3.3 Interactive Simulation

Probably the most entertaining command is "simulate." The "simulate" command uses the UNIX fork facility to execute a switch-level simulator called "Son Of ISIM" (SOISIM) [Szym85]. Icon interacts with the simulator via UNIX pipes and displays the results on the graphic screen. This relationship is shown below:

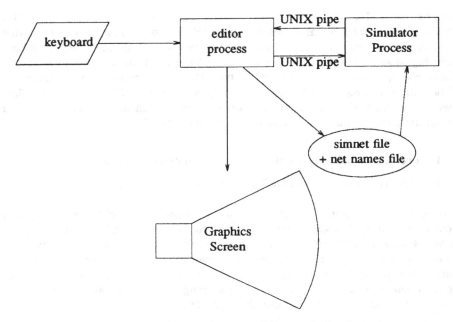

Figure 3-7: Relationship Between Editor and Simulator.

Interactive simulation is supported in both layout and sticks/schematic mode. In the former, solid rectangles represent logic 1, hollow rectangles represent logic 0 and hollow rectangles with an X in them represent logic X. However, today most designers prefer to simulate in the sticks or schematic mode. The color scheme uses red for logic 1, green for logic 0, blue for unknown values, and black for nodes that have been discarded during the compilation phase of the switch-level simulator. During the initial stages of design, debugging consists largely of getting rid of the nodes with unknown value. An example of this is shown in Figures 3-8a and 3-8b.

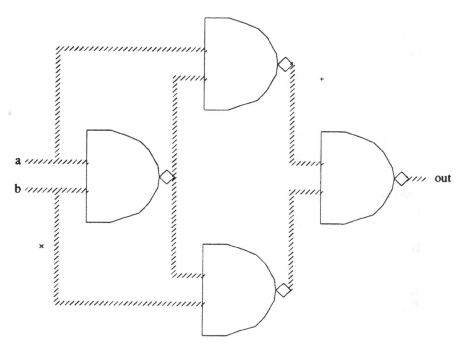

Figure 3-8a: Schematic Simulation Startup.
(hashed lines are unknown, solid are logic 1, hollow are logic 0).

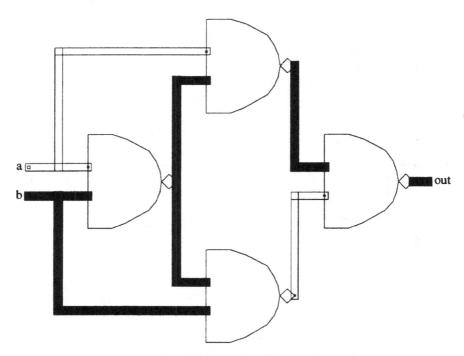

Figure 3-8b: Schematic Simulation Running.

The mechanism supporting interactive simulation is as follows. While simulation is active, the editor keeps track of the electrical net associated with each rectangle it draws on the screen. This requires additional bookkeeping since such information is normally kept accurate only for the current symbol and not for objects inside symbol instances it refers to. To keep track of these nets the editor determines the total number of nets represented in the design, including nets local to symbol instances. (This is equivalent to flattening the electrical hierarchy.) The editor then allocates an array of records, one for each net. Each record contains the name of the net, its current value (0, 1, X, or discarded) and the head of a linked list of geometric regions. When the screen is refreshed icon looks up the nets associated with each wire, transistor, contact or port, and adds the geometric region (the rectangle) to the list of regions associated with the particular net. When the user or the simulator changes the value of the net this list is traversed and the color or drawing style of each rectangle is changed. This relationship is shown in Figure 3-9 below:

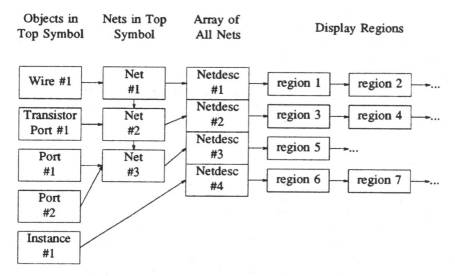

Figure 3-9: Data Structures Supporting Interactive Graphic Simulation.

The bookkeeping gets a little more complicated when the editor has to draw the internals of a symbol instance. To draw the nets correctly the editor maintains auxiliary data structures to support graphical entry of simulation control. Part of this information is a stack of arrays that is generated as the screen is drawn. The arrays map the local nets in a symbol instance into the globally known net array and adds new elements onto the highest level structure as nets are drawn that are local to symbol instances. Whenever the user wants to change a net by pointing to it icon searches all the regions that it drew on the screen linearly to find if one or more overlaps the cursor. The time required for the linear search has never been a problem since it is lost in the user "think time" and the redrawing time.

In addition to pointing to the nets graphically the user can also work with the simulator in its textual interface. This freezes the graphical screen. When the interaction is done the user can return to the graphical state and display the results.

For large designs or highly sequential designs with long start-up sequences, interactive simulation gets tedious, so designers write "C" language programs that drive the simulation and print out the results. One such C program, called "batsim" for "batch simulation," is of special interest. It is a general purpose simulation interface that accepts test vectors in several formats and works with several simulators, including circuit and timing-level simulators, and compares their results. The editor deals with the simulator at arm's length, not as a compiled-in entity. Programs like "batsim" are easy to integrate into the system.

The simulators merely need a few routines to accept and generate the command and data streams. Furthermore, it is even possible to have different versions of the simulator interface run at different times during an interactive editing session because they are not tied in by compilation.

Experience has shown that the graphical mode of simulation and the program-level interface are complementary. The non-graphical interface is invaluable for testing out complete systems when they are nearly correct and when there are errors users tend to go back to the graphical mode to puzzle them out.

## 3.4  Compaction

The basic difference between virtual-grid and fixed-grid IMAGES is that in the former all points fall onto integer grids, while in the latter they may lie on a fractional-unit spacing. For design entry the most noteworthy distinction is that the four connection points of transistors are fixed at a distance of one grid from the center point of the transistor regardless of the actual dimensions of the device. This speeds up editing greatly since everything automatically snaps into place, yet still gives the user some idea how big the layout will be. Once the design has been entered on the virtual grid it can be audited in several ways: It can be simulated in the logic domain with the simulate command and/or with the circuit simulator. (Of course the capacitance values for the circuit simulation are only approximations, since the dimensions of the wiring are not accurate.) It can be checked for wires of different electrical nets crossing each other, or multiple contacts that are coincident. Finally, since the virtual grid layout tends to have the same aspect ratio as the final layout, the virtual-grid design can be used to examine the overall floorplan. Once the design is topologically correct, the compacter can be used to produce a layout free of design-rule violations.

Some designers need to go back and forth between the virtual and fixed-grid designs many times to produce an acceptable layout. That is why a relatively simple compacter is built into the editor as a command: the turnaround can be much faster since the design need not be written out and read in each time. Typically, compaction is used on cells with a couple of hundred transistors, and it takes about 20 seconds (elapsed time) after the compact command is issued before the fixed-grid design starts appearing on the screen. Alternatively, icon supports interaction with the constraint-based compacter MACS. This is somewhat slower, but produces a more compact layout. N.B. A virtual-grid input file does not imply that a classic virtual-grid compacter needs to be used. An example of the output of the virtual grid compacter is shown in Figure 3-10.

ina      inb                                              out

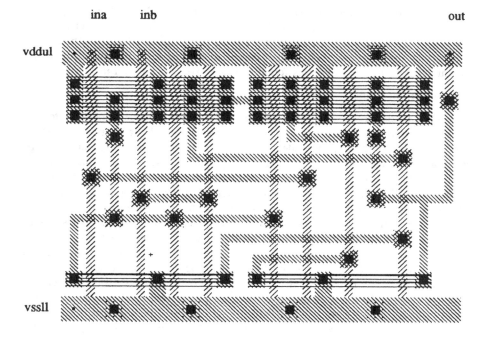

Figure 3-10: XOR Gate After Compaction.

## 3.5 Reading from the Top-Down

Although IMAGES was designed to be as a single-pass, "declare-before-use" language, it is possible to bypass the linear scanning of a large design and read it from the topmost symbol down. This facility is available when the lower cells have been constraint-resolved, and their ports are in known positions. Such is the case on symbols generated by the editor and those generated by the compacter. Any arbitrary cell can be converted into this canonical format by passing through the IMAGES constraint resolver once. Details on this are found in chapter 2.

One consequence of this facility is that editor start-up times are greatly reduced. If a designer wishes to view the overall floorplan of the chip, he or she enters the editor at the highest level. Assuming that there is one file per symbol, the editor starts by reading just the first few lines of the first symbol. A "USES" line specifies the symbols that have instances in the topmost symbol. The compiler then reads just the headers of these symbols. Finished with that, the compiler reads the remainder of the top level symbol and the editor begins drawing. Of course, it cannot draw the internals of the symbol instances yet, but often the designer does not care about them: He or she may only be concerned about their interaction and interconnection. If, however, a "plot all levels" command is issued, the editor repeats the reading process for the visible symbols. That is, it reads their header to find the symbols they refer to, reads the header of the referenced symbols and then reads the remainder of the top-level symbol. This means that the user can zoom into a corner of the top-level design, issue a plot command and just those symbols found in that corner will be read in.

## 3.6  Editor Internals

The editor is the largest single program in the IDA system since it incorporates the compilers, compacter, extractor, output writing code, simulator interface and other facilities. To modularize the code somewhat, it is split into several large groups including: language support, compaction, graphics, and symbol manipulation. Each group can be compiled and tested independently, and many are further subdivided. For example, the graphics interface is split along an interface that separates the device dependent part from the device independent part. This interface consists of a data structure that contains pointers to graphics driver functions. This structure is populated when the editor begins execution and it is determined what type of terminal is in use. For some types of terminals these functions execute locally. For example, the code to support local high-end graphics terminal is implemented by linking it into the editor's executable image. When a command such as "add wire" is issued, it is translated into a subroutine call (such as "draw_rectangle()"), which invokes the UNIX kernel and does a hardware write to the appropriate device. Other terminals are supported by software that runs both on the host and on the terminal in a cooperative mode. For example, when driving a remote workstation, the host sends a stream of ASCII characters on the standard output that encode low level commands (such as the command "draw_rectangle"). The clean interface makes adding new terminals easy.* This function vector interface has the disadvantage of constraining the editor/terminal interface to a predefined set of commands. But it has the advantage of providing a consistent interface when executing locally or remotely. It also provides some performance improvement: It seems that the editor can draw roughly twice as fast when one workstation is traversing the IMAGES data structures and doing the clipping, while the other is drawing the pixels. This parallel execution structure is the type adopted by X-windows for many of the same reasons. In the long run, it is likely that X-windows will become the standard way of using the editor on user programmable workstations, although the other interfaces will still be needed to support non-programmable terminals.

---

* For example, to produce this book, a "pic" graphics capability was added, so that each time line or rectangle was requested, the appropriate pic commands were generated in an output file.

### 3.7 Moving on - Icon as a Non-graphical Tool

When the editor was first developed the primary method of chip design was to create layout by pushing colored rectangles around a screen. For this type of work icon has served fairly well: Its interface is not as slick as some of the contemporary mouse-based editors, but this is compensated by its tie in to the semantics of VLSI. Over the years however, this type of full custom work has become much less important. Rather, designers are spending less time editing rectangles and more time editing designs at higher levels. Circuits are captured as schematics and transformed into layouts using automatic cell generation tools such as SC2 and SC2D. And today, ideas are captured as algorithmic descriptions instead of schematics and turned into circuits using tools like CONES and SLICC. Even higher level layout work such as floorplanning and global writing is done algorithmically. Under these circumstances icon is not used for editing a design, but just for displaying it.

Icon seems to fill that role satisfactorily. It is tuned to short editor start-up times and has a screen format flexible to user needs. In particular, designs have found it handy to decorate layouts with "icons" as a way to back-annotate their floorplans. Furthermore, its graphical simulator interface is useful no matter of how a design was originally created.

But in the long run, even this level of designer interaction seems destined to diminish. Designers dealing with hundreds of thousands of transistors cannot spend time editing detailed schematics, let alone geometry. However, editors like icon will still have a role in this environment as non-graphical command-script interpreters. Their input language is in effect, a hardware description language, and can be used to complement the IMAGES language where it is easier to put a function in one tool than in the language compiler. Today most of the editor scripts are written directly by designers; soon they will be written indirectly by other tools. A typical script in use today is shown in Algorithm 3-2.

```
flatten the hierarchy;
change the width of the power bus to 3 times the default;
extract nets;
add tubs;
check design rules;
rename the symbol;
prepare a circuit simulation file
write
```

Algorithm 3-2: A Procedure of Commands to the Editor.

These scripts provide access to basic design manipulation algorithms in a noninteractive mode, which means the designer is not tied to the terminal. The next chapter discusses some of these algorithms in more detail.

# Chapter 4
# Geometric Algorithms

At the core of most CAD systems are basic algorithms for interpreting and manipulating geometric figures. In some systems, these geometric shapes do not come with a prescribed meaning: for example, in CAESAR [Oust81] the user simply paints regions of the screen with various colors. It is up to the software to find the transistors. In a system based on a higher level language like IMAGES the tools work with more sophisticated primitives. This has several implications: the software tends to be more complex because it has to understand more categories of objects, but it can also tie the algorithms more closely to the original design intent.

Operating between the objects in IMAGES and the shapes seen on a color screen are the algorithms that verify and manipulate the geometry of IMAGES files. These have to tie together the two worlds. This chapter sketches out some of the methods used. It does not provide a complete description but it provides insight into some of the critical ones. Many of the methods use the traditional scan-line algorithm in one way or another.

## 4.1 Enhancing the Basic IMAGES Data Structures

The basic IDA data structures mirror the IMAGES language. For each element in the language there is an object in the data structures, with extra fields to support the common operations. In addition, large complex data structures sometimes exist temporarily in parallel to the basic structures during processing. Usually, these structures point to the objects in the basic data structures. They are allocated as needed and reclaimed afterward. The basic structures are, by and large, not affected by them. There are, however, a few fields in the basic structures that do not mirror the IMAGES language but were provided just to expedite processing. These include two flag bits on each object, named the "mark" and "check" bits. In addition to the "mark" and "check" bits, nets provide the "connected" field that is a pointer to another net.

The algorithms use a special net called the NULLNET to indicate that no net has been assigned to a given object. This is handy in the editor: when objects are created they are assigned the NULLNET, thus avoiding the need to perform an extraction or geometric search each time that the user adds, deletes or moves a group of objects. Thus the presence of NULLNET indicates that a symbol has not been extracted. Many of the other tools such as the compacter, design rule checker, or mask converter check for the existence of the NULLNET on any object and flag it as an error.

A discussion of the basic structures is contained in Appendix A.

### 4.2 Scan-Line Technique

The algorithms discussed in this chapter use a scan-line technique [Bair78] to determine the set of intersecting objects in a symbol. A scan-line can be thought of as a window that hides everything but a narrow vertical strip. This window is wiped from left to right across a design, and as it moves, it discovers new elements as they enter from the right, and forgets about elements as they exit to the left. At any given moment, only a small fraction of the objects in the layout are visible in the window. If the elements are evenly distributed, this will be proportional to the length of one side or about the square root of the total number of elements. It is precisely this reduction in the number of elements that need to be considered at one time that makes the scan-line technique efficient.

### Scan-Line Static Data Structures

To the extent that the designs consist of Manhattan geometry, they can be described as a set of left and right edges of rectangles.* The top and bottom edges can be rederived from them. In these algorithms, the left and right edges are referred to as each other's *mates*. Each edge contains a pointer to its mate as well as a pointer to the original IMAGES object that it represents, such as a wire or a contact.

The edges are sorted in the dominant direction of the symbol. That is, if a symbol's aspect ratio is long in $y$ and short in $x$ then the edges are sorted in the $y$ dimension. This improves the efficiency of the scan-line since it further limits the scope of objects that need to be compared. The basic scan-line code will handle either sorting direction, using the appropriate interpretation of fields; *e.g.* when sorting in the $y$ direction the field "left" is set to indicate that the edge is a

---

\* Non-Manhattan geometry is supported in IMAGES, and in most, but not in all the geometric manipulation tools. For simplicity, this chapter deals with Manhattan geometry only.

bottom edge and cleared to indicate that it is a top edge. For simplicity of explanation however, the rest of the chapter shall assume that scanning is from left to right.

During scan-line processing, a list of edges corresponding to the set of objects visible through the narrow window is maintained, and this is called the "active set." In order for the scan-line to work properly it must be able to see simultaneously all the objects that have an edge in common. This implies that edges cannot be discarded until all the relevant edges have been seen. The consequence of this is that whenever two rectangles have an edge in common, the left edges must appear first in the sorted list, followed by any right edges that occur at the same $x$ coordinate (Figure 4-1). If edge A_R appeared in the sorted list before B_L then A_L would be removed from the active set before edge B_L was seen and the intersection of rectangles A and B would be missed.

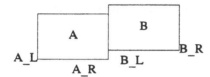

Figure 4-1a: The Edges of Two Abutting Wires.

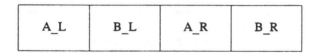

Figure 4-1b: The Sorted List of Edges.

In order for the edge data structures to be accessed efficiently, it is necessary to use pointers inside them that refer to other edges. Yet the set of edges must also be sorted, an operation that normally moves data in memory. To resolve this difficulty an intermediate array is used that has one element for each edge. This intermediate array is sorted and the edge array is untouched, thereby allowing the scanning operation to work through one level of indirection.

The scan-line search proceeds from left to right in the array of sorted edges. Whenever a left edge is found on the list, its object is compared against the set of active objects to see if it intersects any of them in the $y$ coordinate. (The edges store this $y$ coordinate information and level information redundantly with the objects themselves to accelerate this comparison). Any intersections are processed for extraction, design rules, or other purposes, as discussed in the next section, and the edge is then added to the set of active edges. A right edge is processed by removing its mate (which is a left edge) from the active set. To accelerate the removal process, a pointer to the left edge is stored in the right edge, so no search of the active set is required to find and delete the left edge.

## 4.3 Consolidating Rectangles

One of the simpler and more commonly performed operations is to consolidate rectangles. This has several applications in design rule checking, mask generation, and tub insertion. By reducing the number of the rectangles without disturbing the geometry, algorithms run faster and use less memory.

The algorithm used is quite simple. When two intersecting rectangles are found one of the two is expanded through the intersection region if possible (Figure 4-2).

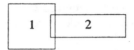

Figure 4-2a: Starting Geometry.

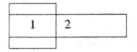

Figure 4-2b: Ending Geometry.

If one rectangle is enclosed by another it is deleted. The consolidating routine is called iteratively until none of the rectangles' right edges are moved. To accomplish this, each rectangle is first put into a canonical form (the lower left and upper right coordinates are correctly ordered)*. A left and right edge is assigned to each rectangle.

The algorithm uses the scan-line technique described above to reduce the number of pairs of rectangles that need to be checked for intersection. Two rectangles, A and B, cannot intersect if the $x$ coordinate of the upper-right corner of A is less than the $x$ coordinate of the lower-left corner of B. Likewise there is no intersection if the $x$ coordinate of the lower-left corner of A is greater then the coordinate of the upper-right corner of B. The $y$ coordinates are be checked similarly. Once it is determined that the rectangles are intersecting another series of four tests on the coordinates are performed to determine which edge is to be expanded. The tests are performed in a binary fashion and the rectangles are

---

\*  The IDA software does not maintain the ordering between endpoints of a wire during editing and
    other operations. So algorithms that depend on this ordering must make a pass that fixes it.

categorized into one of sixteen possibilities.

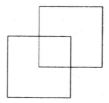

Figure 4-3a: The Corner Overlap Category.

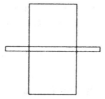

Figure 4-3b: The Crossing Category.

The next step depends on the category. Corner and crossing categories are ignored (Figure 4-3). Some of the categories require that the right edge of a rectangle move further to the right. To track this type of move correctly, the array of edges should be resorted. However, this takes time, and the only loss in not resorting the edges is that some intersections may not be observed. Therefore, an entire pass through the edges is completed before resorting. This has the added benefit that some of the redundant rectangles have now been deleted and thus the second pass works on a smaller set of objects. The routine finishes when a pass completes and no right edges have been moved. This simple geometric merging operation has applications within other operations. These will be described in the next three sections.

### 4.4 Tub Insertion

In CMOS VLSI chips diffusion areas of one polarity must be surrounded by a tub of the opposite polarity. Adding tubs by hand to a complete chip is tedious, non-creative, and potentially error-prone work. This section describes a fast algorithm for automatic tub insertion in a hierarchical design. The tub insertion process is designed to handle a large chip and is therefore feasible as one of the last steps in the CMOS design process.

## Background on Tub Insertion

In all CMOS processes a tub or "well" processing step is used to isolate one or both polarities of diffusion from the silicon substrate. Omission of the tub will produce either a resistor or a diode depending on the polarities of the substrate and diffusion, and is almost always fatal to chip operation. The tub must also be electrically biased. Usually biasing the tub is done by connecting it to Vdd or Vss for n-type tub or p-type tub respectively. At best, a floating tub will disable the enclosed transistors. At worst, it can trigger latch-up, thereby shorting power to ground.

While tubs are necessary for the chip to work, they are not essential to the higher level layout or compaction processes. Logic simulation and net extraction can be carried out without them. Design rules between opposite polarities of diffusion are made such that there is room for the tub to fit without violating any rule. Automatic insertion of tubs on an existing fixed-grid design is therefore a plausible alternative to custom tub geometry provided it can be done quickly and accurately.

One method of ensuring that tubs exist around the diffusion geometry is to perform a "grow" operation on the masks that define diffusion regions. However, this does not guarantee that the tub will intersect a tub contact. Furthermore, it results in a complex mask for the tub with contours that follow every diffusion wire or transistor. Since tub layers are often negatively compensated, notches and small features become a complex problem for something that should be relatively simple.

The preferred approach is to add the tubs by the following algorithm late in the design process. The basic algorithm is as follows: for each piece of diffusion needing a tub, search through the tub contacts and find the closest contact, then insert a tub rectangle to enclose the two. Unfortunately, a naive encoding of the algorithm runs as $O(n^2)$ and is not suitable for large chips. Another problem is that it produces a large number of rectangles, many of which are redundant. Our goal is a fast algorithm that will produce the fewest tub rectangles in the output.

To do this, the inserter performs the following algorithm (4-1).

1.  Build the set of the diffusion areas that are in need of a tub.

2.  Build the set of all tub contacts.

3.  Consolidate small boxes of overlapping diffusion into larger boxes.

4.  For each diffusion area, find the closest contact and insert a tub rectangle to enclose the contact and the diffusion.

5.  Consolidate boxes of overlapping tubs.

Algorithm 4-1: Simple Tub Insertion Algorithm.

The algorithms for each of these steps will be discussed below.

**Building the Sets of Objects**

One of the ways designers squeeze area is to utilize a tub contact in one symbol to bias the tubs in an adjoining symbol. To support this, the tub inserter must expand any hierarchy and work on a flattened description of the design. However, the tub inserter does not require that the whole design be flattened since only the diffusion rectangles and the tub contacts are of interest. The tub inserter keeps track of the diffusion areas, tub contacts, and any existing tubs on three linked lists. It processes the design hierarchically, and for each IMAGES object containing diffusion (wires/boxes, contacts and devices Figure 4-4) its absolute position is calculated and a new record is added to the appropriate list. Tub contacts themselves are added to both the contact list and the diffusion list to ensure that a singleton tub contact will be enclosed by a tub.

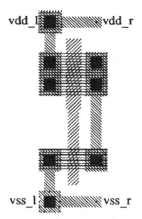

Figure 4-4a: A CMOS Inverter.

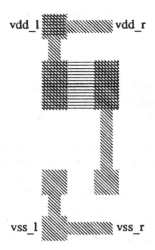

Figure 4-4b: Inverter Diffusion Areas Needing a Tub (metal is shown for context).

Any tubs present in the input are added to the list of generated tubs so that they can be consolidated in the end. The lists are used for the remainder of the process and then reclaimed.

After the list of diffusion rectangles is produced it is reduced with the rectangle consolidating algorithm, resulting in fewer searches by the rest of the program.

**Closest Contact**

One way to find the closest contact to each diffusion region is to check every contact. While this method is quick enough to process layouts involving several hundred devices, it breaks down when the number of devices reaches a few thousand. To accelerate it, a binary tree (a variation of Quad Trees [Fink74]) is constructed to support a branch-and-bound search for the nearest points. The leaves of the tree are the contacts. The tree is built by recursively sorting the contacts into either left-and-right ($x$) or up-and-down ($y$) regions. Each non-leaf node points to two branches: One represents the contacts in the left or bottom side, the other represents the contacts in the right or top side. The non-leaf nodes have bounding boxes associated with them that enclose the contacts in their respective branches. Figure 4-5 shows an example of how four contacts are stored in the tree. At each point, the left subtree contains the contacts with coordinates less than or equal to the node's coordinate, and the right branch has the contacts that are greater than the node's coordinate.

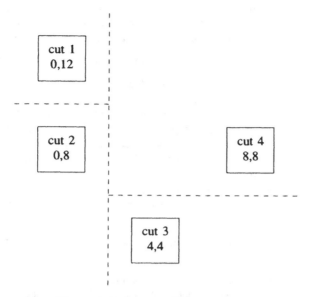

Figure 4-5a: Tub Contact Positions.

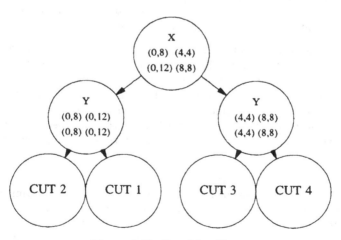

Figure 4-5b: Resulting Tree.

To build the tree, the array of contacts is first sorted in its longest dimension, $x$ or $y$, and the median element used to determine the $x$ (or $y$) coordinate of the root node. Care must be taken when there are several contacts with the same $x$ (or $y$) coordinate: they must all be placed on the left sub-tree for consistency. A node is allocated, its dimension field is set to $x$ (or $y$), and its coordinate is set. The two halves of the array are then processed recursively and the root node is made to point to them. This proceeds until all the contacts are inserted in the tree.

The procedure outlined in Algorithm 4-2 shows how the binary tree is searched.

> Choose one tub contact, and determine its distance. This
> is the initial bound.
> Search the entire tree from the root down, but use the
> bound to help prune branches. That is, examining both
> sides of each branch except when it can be proven that
> any contacts beneath the branch must be further away
> than the known bound.
> Whenever a contact is found that is closer than the existing
> bound, accept it and reduce the bound appropriately.

Algorithm 4-2: The Procedure for Searching the Binary Tree.


The pruning is based on the fact that any contacts on the right side of the node must have coordinates greater in value than the node itself, and those on the left side must have coordinates less than the node. The coordinates of the node determine the position of the closest possible contacts under the node in the tree. If the closest possible tub contact under a branch is further than the currently winning tub contact, there is no need to search that branch.

As in any branch-and-bound algorithm, the efficiency is determined by how quickly the unwanted nodes can be pruned off from further consideration. If only a crude, high estimate is available, then many nodes must be examined because they may contain a contact closer than the estimate. To speed things up, we observe that the diffusion rectangles were originally sorted in $x$, and that it is likely that adjacent diffusion rectangles will be serviced by the same or nearby tub contacts. Therefore, we use the last contact found to calculate the lowest cost limit.

Once the closest contact is determined, an enclosing tub rectangle is created and added to the tub list. After all the tub rectangles have been determined they are consolidated. The consolidation of the tub rectangles reduces the volume of output. More importantly, it reduces the incidence of "pull-aparts." A *pull-apart* is an artifact of the mask compensation process. To adjust for manufacturing changes, each mask rectangle is compensated (grown or shrunk) to produce what is asked for in the design. Sometimes the compensations are negative, because the fabrication process results in a growth beyond the drawn shape. A negative compensation results in a rectangle being shrunk in both dimensions. If two rectangles barely overlap and are compensated negatively a small gap between the rectangles will be formed and appear as an anti-feature on the mask (Figure 4-6).

The nature of the consolidating algorithm expands the barely overlapping rectangles to fully overlapping rectangles, thereby reducing the number of possible pull-aparts.

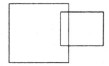

Figure 4-6a: Likely Pull-Apart.

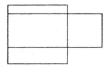

Figure 4-6b: Consolidated Rectangles. No Pull-Apart.

On the other hand, an even simpler approach is often desirable. While the tub inserter is designed to work on an entire chip it can also be run on individual symbols. On a gate-matrix design, such as one produced by SC2 as discussed in Chapter 5, the tub can be one large rectangle encompassing the entire diffusion of the proper polarity without interference from the opposite polarity diffusion (Figure 4-7). This is the form that human designers usually create and they expect as much from a tool. A switch is provided on the algorithm that takes the minimum and maximum $x$ and $y$ coordinates of the diffusion needing tub and the tub contacts. One large tub rectangle is produced to cover the rectangle. The algorithm is fast, it creates just one extra object, and handles a broad set of users.

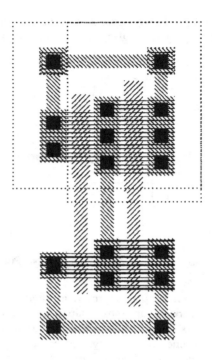

Figure 4-7a: Two Tub Rectangles.

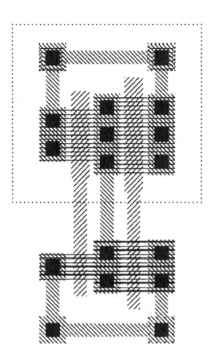

Figure 4-7b: One Large Tub Rectangle.

Unfortunately, the basic algorithm outlined here has the flaw that it can potentially create collisions with the other diffusion polarity. If a piece of diffusion has a piece of the opposite polarity diffusion between itself and the nearest tub-tie, an electrical short will result. However, this has not proven to be a problem in practice since such a glitch is caught quickly by the design rule checker. Designers seem to find this policy is a feature, not a bug, since it quickly catches points with inadequate tub ties. Whenever it occurs (which is very rare), the best cure has always been additional tub contacts closer to the diffusion that needs the tub.

## 4.5 Hierarchical Net Extraction

Net extraction is the process of inferring an electrical model from geometry. Because of the likelihood of human error, verification is always required for hand edited designs. Similarly, because of the possibility of software errors automatically generated layouts are also normally verified. (It sometimes seems that VLSI CAD tools fall into two categories: the buggy modern ones and the stable obsolete ones.) Layout verification can be divided into two parts: circuit extraction and design-rule verification. At a minimum, circuit extraction is the process of deriving a transistor connectivity list from the geometric information. More elaborate circuit extraction models parasitic circuit elements, especially routing capacitance, and perhaps even resistance and inductance.

Like most extractors, the extractor described here uses a scan-line technique as its basic geometric engine. Unlike most extractors, the IDA extractor operates hierarchically, which can significantly reduce its time and storage requirements. The net extractor assigns every physical object an electrical net with a meaningful name. The information has two immediate uses: it simplifies design rule checking (done here as a separate step) and it allows a transistor-level model of the circuit to be generated, perhaps augmented with parasitic capacitance information. By decoupling the basic extraction operation from the detailed design-rule checks the designer can take advantage of the symbolic nature of the IMAGES language: the electrical information can be expressed symbolically, and need not always be rederived from the geometry. Indeed, one of the best ways to use this extractor is to verify that the intended electrical model is implemented correctly in the geometry.

The extractor described here is object-based and hierarchical. The word *hierarchical* is critical here. Traditionally, net extraction programs such as HCAP [Swar83], and GOALIE [Szym84] work on a flattened description of a VLSI design. These tools are capable of doing a detailed parasitic extraction. However, because they manipulate a flattened description they require computer storage and time proportional to the total size of the chip. In contrast, the hierarchical method described here usually requires resources proportional to the size of the description file, which is typically much smaller due to the repetition of subsymbols.

## Basic, Non-Hierarchical Extraction Operation

The extraction process is divided into three operations: a setup phase, which allocates and initializes auxiliary data structures, the scan-line phase, and a post-processing phase that consolidates the net information. The set-up phase starts by storing away the name of the net associated with each physical object in a table. This is used at the end of extraction for verification and net naming. The extractor then strips the existing nets off all the physical objects, and reclaims the storage associated with them. A new, unnamed net is allocated for each physical object. The unnamed net's "connected" field is made to point at itself.

To use the scan-line technique, one or more rectangles are allocated for each IMAGES object. These rectangles are back-linked to the basic IMAGES objects that created them. For most objects the breakup is obvious, but for others, there are several possibilities with different advantages and disadvantages. For example, the extractor represents transistors as three rectangles, one for the gate and its overhangs, and one each for the diffusion regions that form the source and drain extensions. No separate rectangle is allocated for the active region. The data structures include a port object for each of the four connection points on a transistor; they are named "src," "drn," "gt1" and "gt2" (Figure 4-8).

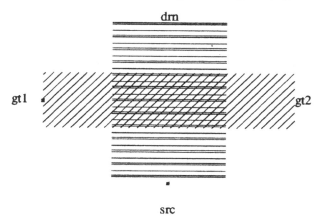

src

Figure 4-8: Three Net Areas of the IMAGES Device.

During extraction, the edges used to represent a transistor's source and drain point at the "src" and "drn" objects on the basic IMAGES data structure, and the gate region points to "gt1." Since both of the polysilicon gate overhang regions point to the same object automatically implies that they represent the same electrical net.

Contacts are handled by allocating a pair of edges for each routing level in the contact, and having them all point back to the original contact data structure. Again, this implies that they are electrically connected. In contrast, each port is given its own net; their default representation is a rectangle of zero width and zero height, modeled using two edges.

Scanning proceeds from left to right, or bottom to top, depending on the aspect ratio of the symbol. As intersecting edges are found it is necessary to use a "union-find" algorithm on their associated nets. Rather than the traditional "forest of trees" method [Sedg83], this algorithm is implemented here using a circular list held together with the "connected" field included with each net (Figure 4-9). When two edges are found to intersect, their net lists are made into one list by adjusting the list pointers. Care must be taken not to insert a net into the list twice: this is prevented by assigning nets an arbitrary ranking and checking the rank relationship before insertion. It is this ranking that makes the algorithm efficient.

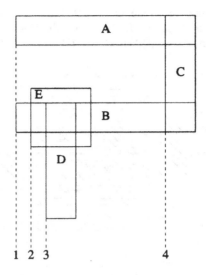

1.      A.net.connected = A.net
        B.net.connected = B.net

2.      A.net.connected = A.net
        B.net.connected = E.net
        E.net.connected = B.net

3.      A.net.connected = A.net
        B.net.connected = E.net
        E.net.connected = D.net
        D.net.connected = B.net

4.      A.net.connected = C.net
        C.net.connected = B.net
        B.net.connected = E.net
        E.net.connected = D.net
        D.net.connected = A.net

Figure 4-9: The Value of the ''Connected'' Field at each Scan-line Location.

After all the edges are processed the connected nets form a linked list (Figure 4-10a).

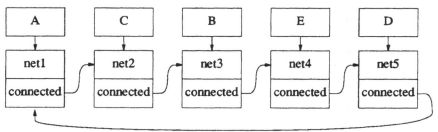

Figure 4-10a: The Linked List Formed by the "Connected" Field.

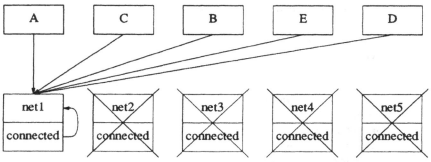

Figure 4-10b: The Nets After Consolidation.

These lists typically contain a couple of dozen elements corresponding to the elements in the circuit that belong to the same net. Before the extractor is finished, however, these lists must be reduced to just one net. This is done in linear time by using the connected field again.

The following procedure (Algorithm 4-3) is used to gather the nets together after scanning. It makes all the objects whose nets are linked point to the same net, and marks the extraneous nets for garbage collection. The result is shown in Figure 4-10b.

```
reduce_nets()
   for each net n
       n.check = false
   end for
   for each net n
     if !n.check and !n.connected.check then
       n.check = true
       tmp_net = n.connected
        while tmp_net != net
          tmp_connected_net = tmp_net.connected
          tmp_net.connected = n
          tmp_net = tmp_connected_net
        end while
       n.connected = n
     end if
   end for
   for each object o
     o.net = o.net.connected
   end for
   for each net n
     if !net.check then
        reclaim_storage(n)
     end if
   end for
end reduce_nets
```

Algorithm 4-3: Gathering and Reclaiming Linked Nets.

## Adding Hierarchy to the Extractor

As discussed above the extractor does not just manipulate polygons, but retains pointers to the original IMAGES object primitives. These can include instances of subsymbols, and in particular the ports on these instances ("inst-ports"). Information about electrical nets is passed from a symbol's definition up the hierarchy to the place where the instance appears. These ports form an abstract view of the symbol: no other information about the internals of the symbol is visible from the outside.

Instance ports are the key to hierarchical net extraction. Since the inst-ports describe the electrical connectivity of the called symbol's ports, they can be used to determine if the two ports are connected together in the lower symbol. For example, in Figure 4-11 a base symbol has four ports but only three externally visible nets.

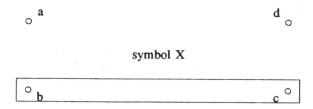

Figure 4-11a: The Internals of Symbol X.

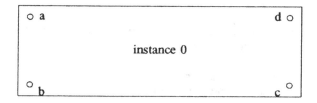

Figure 4-11b: A View of the Instantiated Symbol X.

In this example ports "a" and "d" point to different nets, but ports "b" and "c" point to the same net.

The relation among the nets associated with inst-ports is directly reflected in the edge data structures and corresponds to the nets in the defining symbol. This association is built before the scanning, using cross-symbol net pointers.

The procedure set_inst_port_nets given in Algorithm 4-4 will assign the inst-ports of a given symbol to a new net. The results of the above algorithm can be shown graphically in Figure 4-12.

```
set_inst_port_nets()
  for each instance i

    /* Each base symbol may be instantiated more than once */
    /* i.definition is the symbol that defines instance i */
    for each port p in i.definition
      p.net.check = false
    end for

    for j in 0 .. i.definition.num_ports
      /* ip is matched with dp */
      ip = i.inst_port_j
      dp = i.definition.port_j
      /* if the net of the defining port has not been seen before */
      if !dp.net.check  then
        /* create a new net */
        new_net = create_a_new_net()
        /* assign the new net to the inst-port*/
        ip.net = new_net
        /* assign the connected field of the
        ** inst-port net to point at itself */
        ip.net.connected = new_net
        /* This is a cross-symbol assignment of the net */
        dp.net.connected = new_net
        /* Mark the defining port's net as having been seen */
        dp.net.check = true
      else
        /* Use the cross-symbol assignment to find the
        ** net in the current symbol */
        ip.net = dp.net.connected
        /* Make sure the connected field is correct */
        ip.net.connected = dp.net.connected
      end if
    end for
  end for
end set_inst_port_nets
```

Algorithm 4-4: Associating Nets with Instance-Ports.

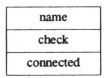

Figure 4-12a: Some Fields of the Net Structure.

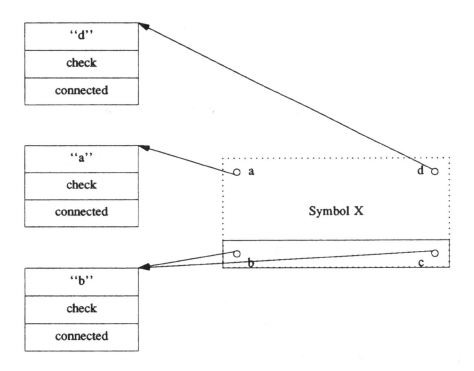

Figure 4-12b: Symbol X with Its Nets.

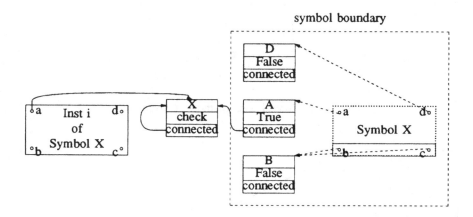

Figure 4-12c: The Net Fields After Inst-port "a" Passes Through the Main Loop.

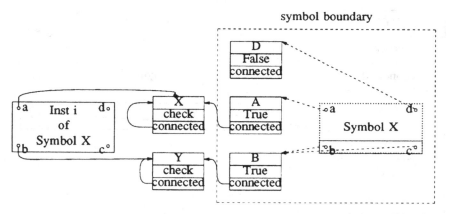

Figure 4-12d: The Net Fields After Inst-port "b" Passes Through the Main Loop.

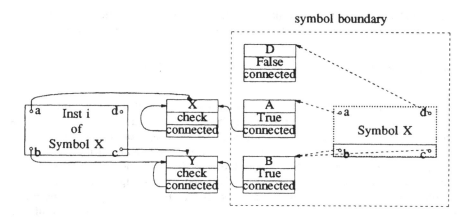

Figure 4-12e: The Net Fields After Inst-port "c" Passes Through the Main Loop.

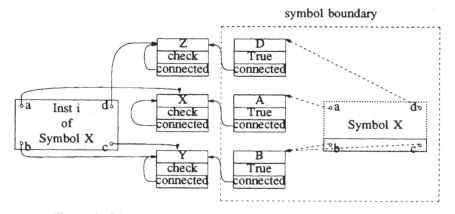

Figure 4-12f: The Net Fields After All the Inst-ports Are Done.

These operations take time proportional to the number of ports in the defining symbol. When they are done, the electrical relationship among the inst-ports is reflected in their net pointers. That is, ports that share an electrical net in the defining symbol are represented by inst-ports that share a net in the calling symbol. These inst-ports are represented geometrically by rectangles of zero height and zero width.

Once the edges are built and their associated nets constructed, the scan-line proceeds from left to right. When edges associated with inst-ports are encountered and they geometrically intersect another object, their nets are connected using the union-find algorithm described above. They may intersect any type of object including other inst-ports. (At the present time, ports and inst-ports have zero height and width, and they can interact with each other only if they are coincident. But future extensions to the IMAGES language may relax this requirement.)

## Design Method Issue - Extraction and Design Rule Checking

This hierarchical method of net extraction will not detect all possible short circuits. For example, a metal wire can run across the top of an instance far from any of its ports, but touching its internals by mistake. If the chip were extracted, it would simulate correctly, but in the final mask this wire would short out the metal wires in the lower level symbol. That is why the extractor is considered to be a tool for extracting "design intent." It must be followed by a design rule checker, that verifies that the implementation matches the intent. In this example the two metal objects with different nets would be found touching each other, which would be flagged as an error. Catching the short circuit in this manner indicates immediately where the problem lies, and provides a useful diagnostic message. Other systems that work without design hierarchy, or that combine extraction and design rule checking, might not catch this, since they could not determine that the nets were not connected intentionally.

## Practical Considerations

In order to cooperate with other design tools, the extractor must obey the rules and maintain the semantics of the design. Not surprisingly, this has led to a substantial increase in its complexity. But this is essential for user acceptance. One issue that proved critical was the choice of net names. The extractor stores initial net names in a table. After extraction, the newly allocated nets get their names from this table wherever possible. However, if a conflict exists between the original and extracted connectivity, a warning is printed and the extractor makes up a new name for each new net.

Some nets may be marked with special properties. In particular, it may be that a local net must be connected to power or ground externally in order for the symbol to work correctly. This is called the "intended net" feature, and it should be preserved across extraction if possible. Unfortunately, the data structure associated with the net is reclaimed early in extraction, and this information can be lost. To preserve it across the extraction operation, it is attached to a port before extraction begins. After processing, the extractor examines the properties on the electrically connected ports and binds them onto the new nets. Any conflicting properties are reported to the user.

New names are also required if the design was created geometrically in the editor and was being extracted for the first time. To facilitate this the extractor first "inverts" the relation between nets and IMAGES objects. That is, all the objects on a net are attached to the net by a linked list. The extractor attempts to create a new net name by trying the following names in order:

1. The intended net name of some port on the net.

2. The intended net name of some mark on the net.

3. The name of some port on the net. If a net by this name already exists in the current symbol some other port is tried until all ports on the net are exhausted.

4. Similarly, any or all "mark" names.

5. Similarly, any or all transistor port names, such as "pullup_src."

6. Similarly, any or all contact names.

7. Finally, unlucky nets that collide with existing names or nets that have only wires on them get a synthetic name such as "net__77."

The inversion of the net lists, and their traversal, requires only time proportional to the number of objects in the symbol. Checking for name collisions is expedited by a hash table.

## Fixed-Grid versus Virtual-Grid Extraction

Because the IMAGES language is used in several modes, the extractor must operate according to several sets of rules. Specifically, fixed-grid designs require the traditional type of extraction where any two objects of the same level that abut or intersect are assigned the same net. On the other hand, virtual-grid designs require a schematic type of extraction where only coincident objects are assigned the same net. The consequence of this is that wires are only connected at their endpoints. The example in Figure 4-13 is a small schematic of a flip-flop based on NAND gates.

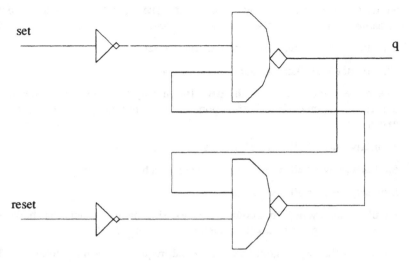

Figure 4-13: A Schematic of a Set-reset Flip Flop.

Under fixed-grid extraction the crossing wires will have the same net, and the circuit would be meaningless. The same design interpreted under virtual-grid extraction rules will not short the wires and the circuit would be correct. If the designer intends that the wires should be connected, he or she should put a mark at their intersection. While the purpose of this style interpretation is clear for schematics, what may not be obvious is why this should also be used on virtual-grid layout cells. The motivation here is early error detection. For example, Figure 4-14 shows a virtual-grid design with a bug in it: two poly wires that cross each other and short out unrelated parts of the circuit. The extractor will mark them as having separate nets. Because the compacter does an audit before compaction for two conflicting nets at the same virtual-grid point, this bug will be caught long before simulation.

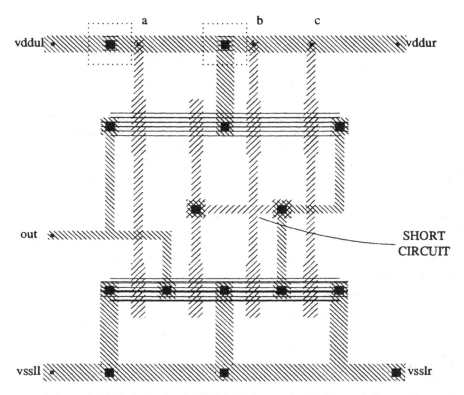

Figure 4-14: A Virtual-grid Cell with Shorted Poly Beneath Port "b".

While this process of hierarchical extraction is designed to be conservative, it does open a window of vulnerability: To extract a higher level symbol correctly it is imperative that the nets associated with the ports on lower-level symbols be correct prior to extraction. If they are not, the extraction process is meaningless even if the designer subsequently extracts the lower-level symbols. To guard against this possibility, it is possible to use modification dates on files, and/or labels on IMAGES symbols (such as "DIRECTIVE extracted at DAY XX time YY;"). Such a system could even automatically re-extract any symbols in doubt. However, no such system is currently in use because the extractor is so fast that it is simpler to just re-extract all symbols from the leaf cells up if there is any doubt about their currency.

**Uses**

The circuit extractor can be used in several ways during chip design. If a design is entered graphically, the extractor can be used to determine the initial electrical connectivity. In this case, it is attempting to derive design intent. The key issues here are usually the selection of meaningful names, and the distinction between correct virtual-grid designs and layout errors. After a design is in the system, the extractor is used more as a design verification tool: it will post notice if the net associated with a geometric object has been changed. Typically, since only a few leaf cells are created in the editor and the rest created using layout generators or the IMAGES language directly, design verification is a more frequent operation than the capture of design intent.

The time and memory efficiency of the extractor is based on two factors: the speed of the underlying scan-line operation, and the (greater) gain of the hierarchical extraction strategy. The combination of these two techniques has been wholly satisfactory: most users find that extraction is so fast that it is now in no way a limiting factor to chip design.

## 4.6 Hierarchical Design Rule Checking

Like extraction, design rule checking is a classic problem in CAD. In many systems the two operations are combined into one general geometric manipulation engine. But here the design rule checker is an object-oriented program that makes use of the symbolic nature of the IMAGES language and the associated electrical net information. The electrical net information must be available before it begins. Rather than striving for complete generality, the design rule checker is optimized to popular MOS processes. The rules it supports come in two types, the conventional minimum distances and the minimum gaps. These minimum gap rules are required to avoid "anti-features" and the design rule checker can generate patches to fill them when necessary.

The current IMAGES design rule checker ("imdrc") was derived from the original "i" language design rule checker ("drc") written by Steve Johnson [John82]. While both tools were alike in that they processed the design hierarchically, the original one had several algorithms that were $O(n^2)$ in the number of boxes and instances in a symbol. These have been largely replaced with scan-line and linear search techniques. These tools not only take advantage of design hierarchy to speed the checking of design rules, but also to reduce the amount of storage required. Other design rule checkers such as LARC2 [Bona86] require disk storage for a flattened description of the chip plus the intermediate form of edge storage. The IMAGES checker, on the other hand, only needs storage proportional to the size of the IMAGES description of the chip.

## Design Rule Table

A rule table is initialized by the program. The rule table contains the levels that interact with a given level via some design rule. This table is used to eliminate extraneous searches (*e.g.* diffusion interacts with poly, but not with metal) in the scan-line searches and the " box-to-instance" and "instance-to-instance" routines described below.

## Scan-Line Modification

The program uses the same basic scan-line procedure as outlined above. Because design rule checking is a frequent operation, the data structures were customized to improve efficiency. IMAGES objects are again broken into the boxes that they represent. This time, the edges used in the scan-line represent the box after being grown by an amount equal to one half the maximum design rule associated with the level of the box. As will be described later, active areas of devices have special properties. In order for the properties to be effective, the active areas are grown by one half their maximum design rule plus one half of the maximum design rule of the diffusion material in the device. The grown edges are sorted in order of ascending $x$-coordinate, with a right edge before a left edge of the same $x$-coordinate. This ordering of edges has the property that two boxes need to be checked for a design rule violation only if the right edge of the left box occurs after the left edge of the right box. The half-rule growth assures this: if the right edge of the left object is encounted before the left edge of the right object, the two objects must be separated by at least the maximum design rule.

To accelerate processing, the active set is more complex than the one used for extraction. It takes the form of an array of linked lists indexed by level. As an edge enters from the left, its level is looked up in the rule table. It is only necessary to check the levels appearing in the table for design rule violations rather than everything in the active set.

## Hierarchical Mapping of Nets

Another increase in complexity comes from the bookkeeping necessary to keep track of nets across the symbol hierarchy. In the extractor, information is passed in one direction: from a symbol definition to the place where it was instantiated. In the design rule checker, it passes in both directions: Electrical nets are mapped from enclosing symbol to instantiations using a mapping of the port nets to the instance ports nets (Figure 4-15). The hierarchical mapping of nets allows the design rule checker to know whether two given boxes are electrically connected or not. This is true independently of where the two boxes occur relative to each other in the hierarchy.

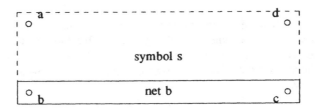

Figure 4-15a: The Internals of Symbol "s".
Net "b" is connected to the outside through ports "b" & "c"

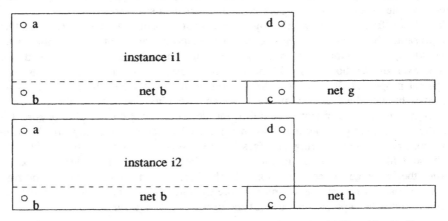

Figure 4-15b: Instance "i1" of Symbol "s" Maps Net "b" to Net "g".
of the Containing Symbol. Instance "i2" of Symbol "s"
Maps Net "b" to Net "h" of the Containing Symbol.

**Program Flow**

After the IMAGES file is parsed and its geometrical and electrical constraints are resolved, device sources and drains are geometrically subtracted from diffusion wires of the symbol. Each physical object in the symbol is converted into its respective boxes, and boxes of the same level are merged. The resulting boxes of the merge are grown and then used to populate the edges. Each symbol is then checked using the scan-line technique to detect interactions. Each interaction can be checked with one of the following three functions:

1.  check a box against another box

2.   check a box against a symbol instance

3.   check an instance against another instance

These steps are described in detail below.

### Box-to-box Edge List Construction

The IMAGES language allows the designer to describe wires that extend past their endpoints or turning points by half their width*. It is possible, especially when connecting wide diffusion wires to wide transistors, for the end of the diffusion to extend under the active area of the transistor. This can cause spurious error messages. The solution is to trim these wires so that they extend only to the edge of the active area of the transistor. For simplicity, however, a large wire being connected to a port of an instance is left untrimmed, even though it might extend under the active area of a transistor inside the instance.**

After clipping diffusion wires, boxes of the same level and same net are merged to eliminate extra checking and notches. The boxes are merged using the rectangle consolidating algorithm previously described.

The edges of the symbol are stored in two ways. One is as a sorted array of all the edges in the symbol. The other has each left edge in a linked list for a given level (Figure 4-16). In this structure, left to right scanning can occur on all of the edges or only the edges of a specific level.

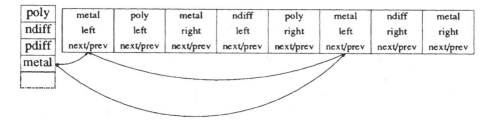

Figure 4-16: Edge Storage Array Linked by Level.

---

\*   This optional feature is useful for wires with many bends in them, since it avoids forming notches in the outside corners at turns.

\*\*   This is one case in which it is possible to configure a test that will cause the design rule checker to report an error where none exists. On the other hand, these cases are rare and there are other ways for the user to describe the same layout so that no error is reported.

All of the physical objects in the symbol, *viz.* instances, wires, contacts and devices, are broken into boxes with appropriate levels. Each edge represents its box grown by half the maximum design rule entered into by the box's level, except for device active areas, which are grown as described above.

An array of bounding boxes indexed by level is calculated. The bounding box of a specific level is used by "box-to-instance" and "instance-to-instance" to expedite searching for intersections. Each bounding box level is calculated based on the grown edges (Figure 4-17). The symbol's overall bounding box is the smallest rectangle that will enclose all of the bounding boxes of the different levels, and it determines the left and right edges of the symbol instance.

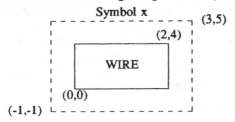

Grown wire represented by the edges

Figure 4-17: Symbol x Has Bounding Box Coordinates (-1,-1) (3,5).

## Box-to-Box

The core of the design rule checker is comparing one box to another. Obviously, this routine has to detect when two boxes are closer than the design rules permit. But it also has to know how to check for small features, and to know when to suppress spurious warnings as discussed in a later section.

In the design rule checker, each box is processed in turn and checked with the box-to-box procedure against all boxes that could be in conflict. The box that is held constant is called the "box under test." The box-to-box procedure accepts the "box under test" and one other box, and calculates the Euclidean distance between them. It then compares this distance to the appropriate rule for the boxes' levels. If there is enough space the procedure returns. When there is not enough space, it goes on to search for a "valid excuse" for the closeness, such as for a device active area that is interposed between the two boxes. In the absence of such an excuse, an error is reported.

## Box-to-Instance

The purpose of the "box-to-instance" operation is to check the box under test with the internals of a symbol instance. It does this by looking inside the definition of the instance, at the boxes that it represents, and comparing them to the box under text by invoking "box-to-box" repetitively. To take advantage of the sorted scan-line edges, the function "box-to-instance" takes the box under test and translates it into the coordinate system of the lower-level symbol (see Figure 4-18). The box is then compared to the edges stored with the original definition of the symbol. The alternative to this might be to take each box in the lower-level symbol and compare it with the box under test in the higher-level symbol. But translating the higher-level box into the lower-level symbol's coordinates saves processing time in two ways. First, there is only one translation done per box-to-instance situation: the alternative would involve flattening the instance and translating each box in it. Second, the sorted edges of the instance's defining symbol can now be reused in their original form in the scan-line algorithm. The scan-line stops searching as soon it hits a left edge that is right of the box under test's right edge. Since the box under test has been rotated and translated, if any errors are to be reported the coordinates in the error message must undergo the reverse transformation.

Figure 4-18a: Symbol "e".

Figure 4-18b: Instance "m" of Symbol "e" Rotated 270 Degrees, and a Box.

Figure 4-18c: Box Rotated into Symbol "e's" Coordinate System.

The "box-to-instance" routine is optimized to look only at boxes in the defining symbol whose level is related to the box under test. The program expedites this in two ways. First, it uses a technology derived table to determine what levels the box under test must be checked against. In addition to the scan-line array, each symbol stores bounding-box information for each level such as metal, poly, ptub, *etc.* (In many cases, these bounding boxes are empty: *e.g.* a routing symbol may have no diffusion material in it at all.) The box under test is compared to each of the bounding boxes of the relevant levels in turn, and if no intersection is found, the program goes on to the next relevant level. The second optimization makes use of the double linking of the edges in the scan line. If the box under test intersects a level's bounding box, the design rule checker tests for violations against all the relevant edges, but in doing so skips through the edges using the links. Only the edges whose levels are related to the level of the box under test are examined for a violation.

A complication arises due to the hierarchical nature of IMAGES: the instance being examined may have another symbol instance within it. If box-to-instance finds another instance inside the instance it is examining, it makes a copy of the second instance and transforms the copy into the original symbol's coordinate system. This entails subtracting the translation and rotation of the first-level instance. For example, the high-level symbol in Figure 4-19c is made up of an instance of symbol "m" (shown in Figure 4-19a and 4-19b) and a box. While symbol "m" is being examined by "box-to-instance" it finds an instance of symbol "e." A new, temporary instance is created in the original symbol at a position that preserves the relation between it and the box under test.

Figure 4-19a: Symbol e.

Figure 4-19b: Symbol m Containing Instance of Symbol e Rotated 270 Degrees.

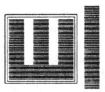

Figure 4-19c: High-level Symbol Containing Rotated Instance of m and a Box.

Figure 4-19d: Relationship as seen by Design Rule Checker
(One level of hierarchy has been removed.)

Box-to-instance then recurses on the untranslated box under test and the new translated copy of the instance. The new copy is added onto a call chain list, so that if any errors are found the user can trace the lineage of the offending boxes to their proper symbol.

To get the correct net information, a copy of the box is made and the net is mapped into the current symbol. When box-to-instance recurses, the net is mapped again.

### Instance-to-Instance

The purpose of instance-to-instance is to look at each of one instance's boxes and compare it to the other instance. Instance-to-instance will always choose to decompose the instance with the least number of edges, because it is likely that many of the edges will not intersect the second instance at all, thus minimizing the number of comparisons required. The decomposed instance is referred to as the "instance-under-test."

As mentioned before, each symbol carries the bounding box associated with each level occurring in it. The first step in instance-to-instance is to look at the technology rules to determine if there exists any level for which the boundary of the instance under test overlaps with the boundary of the other instance. If none exists, their can be no conflict. But if one or more of the bounding boxes conflicts, it checks each box in the instance under test using box-to-instance. For efficiency, the "instance-to-instance" routine handles an instance on the edge list (called a *subinstance*) specially: it can sometimes eliminate all checks with the subinstance by considering bounding boxes. However, if a non-zero overlap is detected on some level then a copy of the subinstance is made by translating it into the original symbol's coordinate system, and handling it as if it were an instance in the original symbol. This involves some amount of bookkeeping to keep the net information straight, and to make error messages sensible to the user.

### Suppressing Extraneous Error Messages

The object-oriented approach of the IMAGES design rule checker has some disadvantages related to secondary design rule considerations. Some of these are due to the way that it keeps distinct source level objects distinct in the scan-line: no geometric operations on regions take place. One result is spurious error messages. Not all of the "close-encounters" detected by the scan-line algorithm represent real violations. Some are exonerated by extenuating circumstances. Such special cases (a.k.a "valid excuses") fall into three categories: transistor related, contact related, and notches. Each of the special cases is denoted by a flag in the technology file that is tied to the particular rule.

In MOS technologies, there is generally a rule prohibiting diffusion from touching polysilicon. However, overlap is allowed when a transistor is intended. (Determining this intention was one of the original motivations for making transistors be explicit objects.) In a geometry-based design rule checker, this can be accomplished by geometrically subtracting the active regions from the diffusion regions. But in an object-oriented checker, such subtractions are impossible. Therefore, whenever a poly-to-diffusion rule is broken, a check is made of the active set for a covering transistor active region. If one is found, the error message is suppressed. Otherwise the error is reported. A similar technique is used to check for "buried contacts" that make intentional ohmic connect between poly and diffusion.

Another example is the minimum separation rule between contact windows. In many technologies, contact window geometry is crucial to yield: the windows must be uniform and separated by a minimum distance. However, it is permissible (if a bit redundant) to place two contacts directly on top of each other, since this will not influence the final mask. In a geometry-based checker, the two contacts would be merged into one with an initial geometric union step. In an object-oriented checker, contact-contact violations must be checked with the overlap flag. This flag specifies that two windows must either be spaced far enough apart to avoid conflict or must overlap exactly.

For the purposes of checking design rules, a transistor has three electrical regions and four physical regions. The four physical regions are the source, drain, active area, and polysilicon overhangs. The three electrical regions are the source, drain and gate. The source and drain are separated by the active area. The gate of the transistor consists of the active area plus the polysilicon overhangs. The active area is assigned the same net as the gate.

These diffusion regions
might be flagged as conflicting

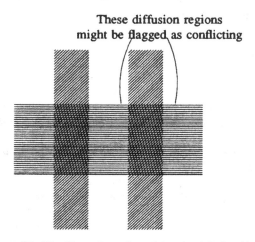

Figure 4-20: The Error Box Saved by the "Trfind" Flag.

This decomposition presents a minor problem in some technologies where the source and drain diffusion areas are closer than the diffusion-to-diffusion rule would normally allow. If, for example, the minimum diffusion-to-diffusion separation is 3.0 microns and the channel length of a transistor is 1.25 microns, then a naive check shows the source to be in violation with the drain. In a geometric design rule checker, one way to handle this is to grow the active region, and subtract it from the diffusion regions. But since subtractions are not convenient in an object-based checker, the IMAGES design rule checker allows the active area of a device to "bless" the interaction of the two diffusion areas that it separates (Figure 4-20). That is, before reporting an error between diffusion boxes, the checker searches for an active region that screens out the violation. On the other hand, the object-oriented approach also has a few advantages around transistors. For example, in some technologies the gate length is smaller than the narrowest poly wire normally allowed. In a geometric checker, this would require geometric steps to determine which poly was part of a wire, and which was part of a transistor. But in the IMAGES design rule checker, this information is available explicitly.

Notches, or anti-features, are defined as edges of the same level and the same electrical net that are closer than the minimum design rule (Figure 4-21a). Such small features cause headaches in the processing line because they cannot be distinguished from flaws in the mask. So it is important to detect such notches before fabrication. Unfortunately, the object-oriented nature of the checker results in spurious error messages: Since its basic search examines only two boxes at a time, it may miss the fact that a third box covers the gap. Therefore, a flag in the rules is used to force the checker to look in the vicinity of the gap to see if indeed another object is present that covers the notch. If the flag is set and an appropriate cover is found, then no error exists.

**Output of the Checker**

The design rule checker produces two categories of output: points where design rule violations have occurred and IMAGES text describing boxes to cover any small features that may have been found (Figure 4-21b). This text is basically appended onto the same file containing the original symbol. In contrast, errors are reported by generating an IMAGES file with boxes overlapping the geometric location of the errors, annotated with a textual explanation. These can be plotted like any IMAGES file, on paper or interactively. The text associated with the error includes the design rule that was violated, the distance between the offending boxes and their electrical nets. Because the error may actually be several levels down in the hierarchy, the error text explains the symbol instance calling chain.

It is possible that an inserted patch will cause the need for another patch. These cases can be handled by running the patched design through the design rule checker again.

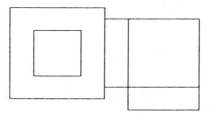

Figure 4-21a: A Notch Formed Between Features of the Same Level.

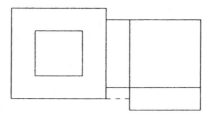

Figure 4-21b: The Patch Inserted to Cover the Notch in Figure 4-21a.


## Potential Enhancements to the Design Rule Checker

There are several speed enhancements possible within the framework of the current design rule checker. Some of these would trade memory for speed. For example, if the scan list were completely doubly linked, the checker could be sped up in some situations by scanning in the right-to-left or top-to-bottom direction. Currently, to check a box near the right side of an instance, the checker has to look at all the edges inside the instance. If it were possible to start at the right side of the instance and work to the left, fewer edges would have to be checked.

Another potential improvement, which has been used for non-object-oriented checkers, would be to break large symbols into horizontal strips, or "hoppers" which would serve to decrease the number of unnecessary comparisons. Currently, the algorithm might compare many boxes that are close to each other in the $x$ direction, but very far apart in the $y$ direction. Hoppers could decrease the number of such comparisons by segregating the boxes of a symbol according to the vertical positions. Each box would belong to every hopper that it intersected, but in many cases this would only be one or two hoppers. Since a wire butting into an instance may only intersect one hopper, only the edges in that hopper need to be searched. Such a system would, however, add complexity onto an already complex tool. Since it is currently not a bottleneck to chip manufacture, these enhancements will be held in reserve.

## 4.7 Summary

This chapter has illustrated how the traditional scan-line algorithms have been applied to the support of the IMAGES language. In each case, a set of auxiliary data-structures have been built to store the edges required for the algorithm, and these have been tied back to the IMAGES objects that spawned them. The alternative is also possible: the edges could be made to stand on their own, and the geometric operations performed without reference back to the defining objects. This has advantages in other systems where the geometry is more "self-defining". But in IMAGES, it is important that error messages and nets be meaningful in the context of the symbols in the original language. Any inefficiency due to this overhead is more than compensated for by the ability to perform operations hierarchically.

# Chapter 5
# Enhancements to the
# IMAGES Language for Synthesis

VLSI layout synthesis requires both general support tools, and tools keyed specifically to the application. To support this the system must be extensible to allow capabilities to be added by CAD workers or the designers themselves (who are sometimes the same person). This chapter illustrates two cases where this has been done. The first cases discusses how the IMAGES compiler and a textual filter can be combined to simulate the effect of having routers in the language. The second section discusses how the IMAGES language can be interleaved with the C programming language so that designers can write technology updatable generators.

## 5.1 Stretchable Routing using IMAGES Constraints

Most VLSI chips consist of relatively few transistors connected by relatively many wires. (Ratios of 1:5 are common.) These wires can be divided into three categories: short wires local to a logic gate, regular busses and random inter-module routing. The IMAGES language, by itself, is well suited to supporting the small wires within cells and the regular busses that are found in arithmetic chips. But it is not particularly well suited to the last.

One traditional argument about VLSI layout has been that well-designed chips do not require routing, because the cells can be made to abut using pitchmatching. This is true only when working on relatively small, homogeneous chips that have a high degree of regularity. For example register/arithmetic units, memories and switching circuits can often be laid out by tiling the plane of the chip with similar cells. However, this model breaks down when dealing with production chips because the control sections of these devices often do not fit into the regular geometry. The layouts of the major modules are related to each other, but independent, and the only practical way to lay out the chip is with routing between the major blocks.

Hand routing for assembling large blocks is tedious and error prone. Therefore, automatic routing of some form is essential in a production CAD environment.

One way to support automatic routing is to provide a separate routing tool. Most routers are of this form, including global routers that assemble large production chips such as LTX2 [Dunl83] These routers, which are sometimes combined with placement software, usually treat objects as rigid rectangles, and place and route them as a final step in the design process. This type of routing is possible with the IMAGES language, and is used on a regular basis. However, it does not make full use of IMAGES: it can be done equally well with any geometry specification language. In particular it does not allow the routing to stretch as object positions are resolved. To do this, IMAGES routing is normally done in a two-pass operation. The routing provided is limited to only channel and river routes, but the integration is unusually flexible.

The routers take advantage of the IMAGES environment in several ways. Their input specification and their output are both in symbolic, constraint-based IMAGES. This makes it easy for human designers and high-level floor-planning tools to read and write routing specification. Likewise, the layout created by the routers is also in a symbolic form, and is therefore able to stretch after routing is completed. The routers may be incorporated in any stage of design: among the transistors and ports in a leaf-level symbol definition, among symbol instances or as a mixture of both. This is in contrast to most routing packages that are used only in a final "layout-freezing" step.

### Background

Even without a set of routers *per se*, the IMAGES language can be a convenient tool for interconnecting subsystems. For example, to connect points out1 through out10 on one symbol to points in1 through in10 on another, a designer might write the following:

```
FOR i in 1 .. 10 LOOP
     WIRE AUTO symleft.outL&(i) RIGHT
      UP TO symtop.in&(i);
     END LOOP;
```

This IMAGES code would produce the results shown in Figure 5-1a.

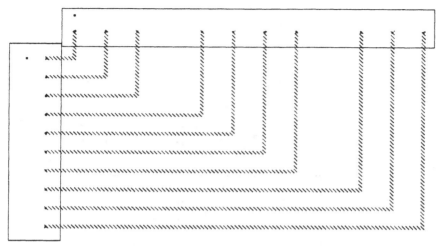

Figure 5-1a: A Simple Route.

This style of chip assembly has several advantages: it is quick, both in terms of human and machine time, it can produce high-quality layout, it is understandable and it is technology-updatable. The last point can be understood by observing the routing contains no numeric positions or displacements. Furthermore, the electrical connectivity is expressed in the same language as the layout. Unfortunately, such routing is easy to write only when the points to be routed cannot destructively interfere with each other. When interference occurs, the designer must resort to more detailed and irregular connection topologies, which can dilute some of the advantages of the symbolic approach. A small corner of a chip done with this method, but on a much larger scale, is shown in Figure 5-1b. This particular chip contains thousands of wires created textually by a designer, who specified the assembly by relative offsets and constraints. In this case most of the wiring was regular, and could be compressed with FOR loops in the input language. But when a small section was needed that had highly irregular connections, the job became tedious and error prone.

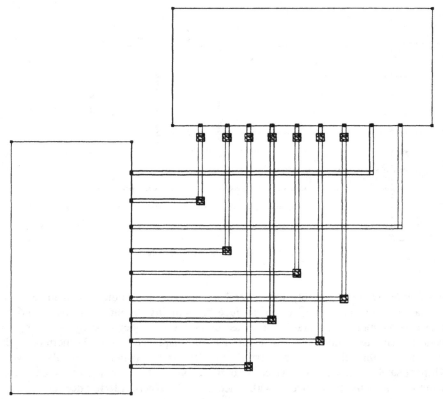

Figure 5-1b: A More Complex Designer-Entered Route.

A set of routers was developed to address these types of complex routes. The routers make the task of routing (almost) as straightforward as the bus-based IMAGES wiring illustrated above. The routers work completely within the constraint-based environment, and provide a quick way to write IMAGES statements for symbol or chip assembly.

## Routing Control Factors

Consider the problem of joining a collection of points on the lower edge of one symbol instance called "upper" with another collection of points near the top edge of a second symbol instance called "lower." In connecting these two symbols, the designer must take several factors into consideration. First, the intended electrical connections between the two symbol instances must be specified. Secondly, floor-planning requirements that specify how the two instances are to be placed relative to each other must be entered. For example, one might want them to line them up so that their ground connections are directly above and below each other. Next, the minimum design-rule spacing from one symbol instance to the other must be entered. This is a function of the internals of the symbol definitions. Fourth, the points that are going to be connected must be specified. Finally, the detailed set of routing wires and contacts necessary that form the actual connection must be created in such a way that they do not come too close to the material inside the instance. The resulting layout must resolve all of these factors simultaneously.

The use of a symbolic, constraint-based environment helps in all of these areas. First, the electrical connectivity can be expressed in the IMAGES language either directly or by implication. For example, a net list can be entered directly using the IMAGES CONNECT statement, as in:

```
CONNECT NET=reset upper.out  lower.in;
CONNECT NET=vss upper.vssll bottom.vssul;
```

or, at a more detailed level,

```
CONNECT pullup.gt1  pulldown.gt2;
```

where pullup and pulldown are names of transistors.

Alternatively, one can create a wire running between symbolic points, and join them electrically with one statement. For example, the statement

```
WIRE POLY pullup.gt2 DOWN TO output DOWN TO pulldown.gt1;
```

lays two segments of a polysilicon wire among three symbolic points, and has the side effect of joining them electrically. In the router interface, both types of electrical connections can be freely mixed. Typically, the connections within symbols are specified implicitly, and explicit CONNECT statements join ports on one instance with ports on other symbol instances. The two techniques complement each other. For example, if there existed multiple, symbolically-specified ground connections inside one instance, connecting to one port would be equivalent to connecting to all ground ports explicitly since the IMAGES compiler recognizes equivalent terminals symbolically.

Once electrical connectivity has been specified, the IMAGES translator can accept the design and it can be simulated to verify its behavior. This can be done before the routing process and independently of it.

The next consideration that must be specified is the floor-planning relationship between the two symbol instances. This relationship should be preserved by routing. The simpliest way is to pin their positions down with a rigid, but relative, offset, *e.g.*:

```
BIND upper.vssll ABOVE bottom.vssul BY 24 ;
```

In most circumstances this type of constraint is too restrictive since the designer does not know how much area the routers will require, nor how the two pieces must be spread to match surrounding circuitry. The solution is to make use of the constraint-resolving features of the IMAGES translator, as in:

```
BIND upper.vssll ABOVE bottom.vssul BY >= 0.001 ;
```

This informs the translator that the top symbol instance must be at least 0.001 units above the lower instance (the number is arbitrary, but must be greater than zero). The keyword "ABOVE" specifies that the ground connections must line up vertically (share the same $x$ coordinate), but are free to move farther apart (in the $y$ direction) as necessary. To account correctly for the design rules between the two symbol instances, an additional constraint can be automatically generated by the PANDA system [Bull86], *e.g.*:

```
BIND upper.vssll ABOVE
  bottom.vssul BY >= 4.75;
```

In the IMAGES environment, it is perfectly reasonable to have both

```
>= 0.001 and >= 4.75
```

constraining the space between the same objects; the translator will produce a solution that satisfies all such constraints simultaneously.

Finally, if the objects to be routed are symbol instances (as is usually the case) the router can consider the bounding boxes of the two objects, and push the two objects apart just far enough that their perimeters do not touch each other or the routing.

The next step is to specify the points actually to be connected. This is done using an order-independent list of symbolic connections, *e.g.* :

```
!channel top
! upper.vssll bottom.vssul upper.out
! upper.reset bottom.strobe bottom.in
!endroute
```

Since the electrical connectivity was specified independently the points can be stated in arbitrary order. There is also no restriction on the reuse of points: A symbolic reference such as `upper.vssll` may appear in several different routings. This form of text is treated as comments by the other IMAGES tools, which means that the file may be compiled and the floorplan examined independently of the routing.

Once all the points to be routed have been specified, the router is run. This router scans the design file and replaces the statements between the `!channel` and `!endroute` keywords with a collection of `WIRE`, `MARK` and `CONTACT` statements. Unlike more conventional routers, it does not produce a rigid symbol containing the routing: The output of the router is almost as flexible as its input. Specifically, all wires and coordinates in the output are positioned by constraining them to either the top or bottom symbol instance. A typical wire path created by the routers would look like:

```
WIRE AUTO upper.vssll DOWN 10
LEFT 5 DOWN >= 5 TO bottom.vssul;
```

In this statement, the `AUTO` keyword instructs the IMAGES translator that the level of the wire is determined by the level of the point `upper.vssll`. The first two segments of this wire (`DOWN 10` and `LEFT 5`) are bound symbolically to the top symbol instance, but the third segment is left unconstrained: The two symbol instances may move further apart if other constraints require it. The utility of this is that symbol-to-symbol spacing will automatically be the minimum necessary to satisfy simultaneously all conditions, *i.e.* the floorplan, design rules and routing-track constraints.

Most channel routing output statements are much more complex than the one above. In particular, a channel route requires contacts that allow wires to change levels inside the route. The router picks a geometric reference point that is attached to one of the sides of the route. When it is necessary to include a coordinate, it is expressed as an offset from that point. The output file created contains declarations of the contacts, with their position constrained to one side. For example:

CONTACT MP zz0_7_9  (row0.g140_tp_341_x,zz0_up_bound) +(0,-28);
WIRE POLY zz0_3_9  VER TO zz0_7_9;
 CONTACT MP zz0_6_11  (row1.g90_tp_37__1_ext_x,zz0_up_bound) +(0,-32);

The choice of the reference point and the calculations of the offset are determined by the directives to the router. This allows the designer to control where vertical slack is to be placed if any is present. This might be important if wires are to be added over the routing region from some other source.

Many independent routes within a symbol definition may be made in one pass. This is the technique used by SC2D, as will be explained in the next chapter. In addition, the router can be run in a second pass on its own output to complete routes that depend on the exact positionings determined by the first pass. This allows a gridwork of routing channels to be created with each pass of the router making the symbol instances more rigid, but still not frozen relative to each other. A small example of this is shown in Figure 5-2.

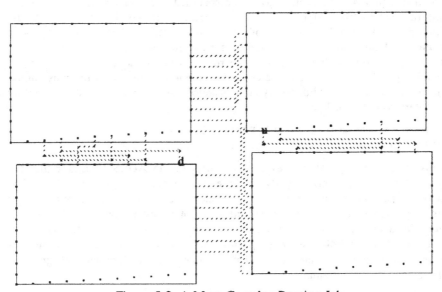

Figure 5-2: A More Complex Routing Job.

## The Router Algorithms

Two categories of routers are provided: a river/ripple router, and a channel/bus router. The first is suitable for connections not involving crossovers. When used in its "ripple" mode it produces wires that follow the next adjacent track as tightly as possible. It can often save area over an ordinary river router, but may add some cross-talk. The algorithm involved is relatively simple: it first calculates the density of the route, and then places the wires in sequence starting on the side that will have the longest straight section.

The channel router is based on a "greedy" algorithm [Rive82] modified to work in a gridless environment with multiple design rules. A "greedy" algorithm is one that looks at only local information in making routing decisions and uses heuristics to decide what to do. The router works from the left side of the channel to the right, and it has been tuned to add jogs at reasonable intervals. It starts off with an estimate of the number of tracks based on the density of the route. A greedy algorithm was chosen because it guarantees completion in the face of an arbitrary number of cycles in the constraint graph. (The constraint graph is the graph that describes restrictions between access on the top and bottom sides of the channel.) It works around cycles by occasionally carrying a signal on two or more parallel tracks, and joins the tracks together when there is space for a vertical jog. It has the unfortunate property that it can take up extra tracks and a large amount of area beyond the right side of a channel. But it has a major advantage in that it is guaranteed to complete the route. Completion is essential to automatic layout systems: Other tools can usually tolerate area inefficiencies, but they cannot tolerate incomplete or incorrect routes. After routing is complete, a "greedy" via eliminator runs on the layout and converts some of the horizontal wires to the same level as the vertical wires if this does not cause conflicts or increase the resistance significantly (e.g. changing any length poly to metal is allowed, but only short metal wires may be changed to poly).

When the routers are used on a fixed-grid IMAGES file they get their rules from a technology database, and understand three spacings: wire-to-wire, wire-to-contact, and contact-to-contact. They automatically determine the specific layers (e.g. metal, polysilicon or n-type diffusion) that are present by examining the list of points to be specified and use this information in calculating the spacings. Additional specifications to the router control such options as via elimination, multi-pass routing and restricted routing areas.

The time required by the routers themselves is usually a few seconds per hundred connections on a one MIP machine and is fast enough that it gets lost in the time to parse the input and write the output (typically about 10 seconds). Routing area of the ripple router is believed to be the minimum necessary; the area of the channel router is dependent on how well its heuristics match the particular layout. In general, however, the router seems to run at the routing density most of the time, and only occasionally take an extra track or two. The heuristics that have the greatest influence on density are the ones controlling the introduction of "dog-legs" into the route. Dog-legs are extra vertical segments that connect routing tracks in a zig-zag pattern. Classical channel routing does not use them, but they are essential to reduce the area of the route.

## 5.2  Impact on Design Methods

The IDA concept of *symbolic* layout is orthogonal to the notion of virtual-grid compaction; symbolic resolution can be done with or without compaction. Because of this, the routers work just as well on a virtual-grid as they do in a gridless environment. (Or better, since the compacter can almost always squeeze a track or two out of a channel route after via elimination.) Moreover, the routers do not require that the points to be routed belong to rigid subsymbols. In fact, the routers can work with any type of object within their scope provided that they can be moved apart if more tracks are needed. The ability to work within a given level is important, because experience has shown that adding symbol definitions that are not re-used elsewhere in the design creates a nontrivial speed and space overhead, and more importantly, obfuscates the intended structural hierarchy of the design.

## 5.3  Designer Written "Generators"

The routers of the previous section are an example of the way a CAD programmer can extend the IMAGES language by linking into the IMAGES data structures after compilation. This is a powerful method since it provides direct access all the technology specifications, design information and algorithmic support libraries. Unfortunately, it also requires considerable sophistication on the part of the developer. Most designers are not interested in this level of programming. They prefer to work in a less complex environment, and the generator writing system is targeted at such designers.

A *generator* is a program that accepts parameters as input and produces a high-quality layout. Generators can be divided into categories according to how tightly they are bound to a particular layout. For example, to categorize a tool, one could ask whether or not it works with more than one set of design rules, and secondly, how broad a set of layouts can be generated. At the tightly bound extreme of these two criteria would be a mask-level layout for one particular circuit. It accepts no parameters and is good only for the original set of design rules. At the most other extreme would be a true silicon compiler, that takes arbitrary design intentions and converts them into any technology from vacuum tubes to gallium arsenide. The layout-based generators described in this section lean toward the tightly bound side of this spectrum. For purposes of discussion, this book makes the distinction between *layout-based* generators (such as a counter generator) and *logic-based* tools (such as standard cell systems). This chapter will discuss the layout-based generators, as a language issue, together with their applications, advantages and limitations. The next chapter discusses logic-based tools.

## Background

Generators have existed in various forms almost from the beginning of VLSI CAD. Probably the first ones were ROM and PLA customizing software that take a list of ones and zeros, and selectively include transistors at particular points in the matrix. Later, more general purpose tools were developed that could generate more flexible, but still highly repetitive layouts, such as RAMs and buffers. One example of such system is the CANONs supplied with the GRED editor [Pott83]. These consist of long sequences of editor commands. Because the editor has decision making and measuring capability some amount of customization is possible. They also have the advantage that the progress of the generator can be viewed directly.

Many systems combine programming languages with layout primitives. Popular base languages include LISP, PASCAL and C, although many others have been used. Each of these systems uses algorithms encoded in the base language to selectively place geometry. Typically, subroutine calls are added that place geometric objects. The IMAGES-C language used in IDA is a descendent of the i-C language described in [Math83]. The advantage of this category of system over languages that work with more restricted geometry description systems (such as CIF) is that more power is available in the underlying geometric environment. Fewer geometric coordinates need to be described since more can be derived from context. Although this requires a more advanced level of training on the part of the user, the amount of effort to produce each generator is lower.

## IMAGES-C Language

IMAGES-C adds to IMAGES the programming flexibility of the programming language C [Kern78]. In an IMAGES-C program a line of text can be either a C statement or IMAGES statement. A simple preprocessor, translates from IMAGES-C to C by recognizing the presence of one or more IMAGES keywords within a line. It leaves C statements unchanged and translates the IMAGES statements into formated C language "fprintf" statements. This allows blocks of existing IMAGES code (say the output of the layout editor) to be incorporated into a C program, which can compute parameters to shape the final circuit. This flow is diagrammed in Figure 5-3. (In this and subsequent diagrams, symbols with double outlines are those that are effected each time the generator is run. Symbols in single outlines are unchanged from one run to the next).

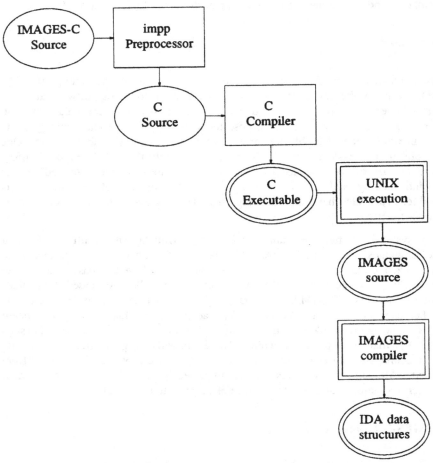

Figure 5-3: Relation of Generator Building Tools.

As an example, consider the following fragment of an up-down counter generator (Figure 5-4a):

```
count(bits) int bits; {
SYM %s{name}_1
IBEGIN
    INST %s{name}cl_1 cntcl;
    FOR i IN 0 .. {bits - 1} BY 2 LOOP
        INST %s{name}cell_1 udcell&(i);
    END LOOP;
    FOR i IN 0 .. {bits - 3} BY 2 LOOP
        BIND udcell&(i+2).mvddul RIGHTOF udcell&(i).mvddul BY %f{pitch};
        WIRE M2 WIDTH=14 udcell&(i).mvddur HOR TO udcell&(i+2).mvddul;
        WIRE M2 WIDTH=10.5 udcell&(i).mgndur HOR TO udcell&(i+2).mgndul;
        WIRE M2 WIDTH=10.5 udcell&(i).mvddmr HOR TO udcell&(i+2).mvddml;
        WIRE M2 WIDTH=10.5 udcell&(i).mgndmr HOR TO udcell&(i+2).mgndml;
        WIRE M2 WIDTH=10.5 udcell&(i).mvddlr HOR TO udcell&(i+2).mvddll;
        WIRE M2 WIDTH=14 udcell&(i).mgndbotr HOR TO udcell&(i+2).mgndbotl;
        WIRE M2 udcell&(i).p2clkr HOR TO udcell&(i+2).p2clkl;
            . . .
    END LOOP;
IEND
}
```

Figure 5-4a: A Fragment of a Simple Generator in IMAGES-C.

The mixture of IMAGES and C is then passed through the IMAGES-C preprocessor to generate a C file, which is then compiled as a program. The result of passing the fragment in Figure 5-4a through the preprocessor is shown in Figure 5-4b.

```
count(bits) int bits;
{
fprintf(icout,"SYM %s_1\n",name);
fprintf(icout,"IBEGIN\n");
fprintf(icout,"   INST %scl_1 cntcl;\n",name);
fprintf(icout,"   FOR i IN 0 .. %d BY 2 LOOP\n",bits - 1);
fprintf(icout,"     INST %scell_1 udcell&(i);\n",name);
fprintf(icout,"   END LOOP;\n");
fprintf(icout,"   FOR i IN 0 .. %d BY 2 LOOP\n",bits - 3);
fprintf(icout,
"BIND udcell&(i+2).mvddul RIGHTOF udcell&(i).mvddul BY %0.3f;\n",pitch);
fprintf(icout,
"WIRE M2 WIDTH=14 udcell&(i).mvddur HOR TO udcell&(i+2).mvddul;\n");
fprintf(icout,
"WIRE M2 WIDTH=10.5 udcell&(i).mgndur HOR TO udcell&(i+2).mgndul;\n");
fprintf(icout,
"WIRE M2 WIDTH=10.5 udcell&(i).mvddmr HOR TO udcell&(i+2).mvddml;\n");
fprintf(icout,
"WIRE M2 WIDTH=10.5 udcell&(i).mgndmr HOR TO udcell&(i+2).mgndml;\n");
fprintf(icout,
"WIRE M2 WIDTH=10.5 udcell&(i).mvddlr HOR TO udcell&(i+2).mvddll;\n");
fprintf(icout,
"WIRE M2 WIDTH=14 udcell&(i).mgndbotr HOR TO udcell&(i+2).mgndbotl;\n");
fprintf(icout,
"WIRE M2 udcell&(i).p2clkr HOR TO udcell&(i+2).p2clkl;\n");
fprintf(icout," END LOOP;\n");
fprintf(icout,"   IEND ;\n");
}
```

Figure 5-4b: Resulting C Language Output.

The C file in Figure 5-4b is compiled with similar modules. The program is run to generate the parameterized layout. The IMAGES language output of the fragment shown in Figures 5-4a and 5-4b is shown below in Figure 5-4c.

```
SYM tmp_1
IBEGIN
    INST tmpcl_1 cntcl;
    FOR i IN 0 .. 3 BY 2 LOOP
        INST tmpcell_1 udcell&(i);
    END LOOP;
    FOR i IN 0 .. 1 BY 2 LOOP
        BIND udcell&(i+2).mvddul
            RIGHTOF udcell&(i).mvddul BY 85.000;
        WIRE M2 WIDTH=14 udcell&(i).mvddur
            HOR TO udcell&(i+2).mvddul;
        WIRE M2 WIDTH=10.5 udcell&(i).mgndur
            HOR TO udcell&(i+2).mgndul;
        WIRE M2 WIDTH=10.5 udcell&(i).mvddmr
            HOR TO udcell&(i+2).mvddml;
        WIRE M2 WIDTH=10.5 udcell&(i).mgndmr
            HOR TO udcell&(i+2).mgndml;
        WIRE M2 WIDTH=10.5 udcell&(i).mvddlr
            HOR TO udcell&(i+2).mvddll;
        WIRE M2 WIDTH=14 udcell&(i).mgndbotr
            HOR TO udcell&(i+2).mgndbotl;
        WIRE M2 udcell&(i).p2clkr
            HOR TO udcell&(i+2).p2clkl;
        WIRE M2 udcell&(i).p2clkbr
            HOR TO udcell&(i+2).p2clkbl;
    END LOOP;
IEND
```

Figure 5-4c: Resulting IMAGES Language Output.

Since the cells of this type of generator are handcrafted before incorporating them into the generator, the resulting layout tends to be compact. However, the compact layout (Figure 5-4d) is only good for the technology in which it was designed. If a new technology emerges the handcrafted cells must be redone and compiled as part of a new executable.

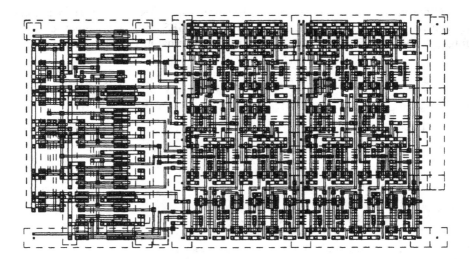

Figure 5-4d: Final Layout.

## 5.4  Design Rule Updatable Generators

Because a generator represents a considerable investment on the part of the designer, there is a strong incentive to try to preserve generators across technology changes. This is not practical for radical changes such as switching to gallium arsenide, but for "shrinks" with similar (but not necessarily linear) design rule changes, the use of the compacter can make the transition less painful.

There are several ways one can build technology updatable generators. One way is to heavily parameterize the IMAGES-C source with many explicit references to technology dependent dimensions. While this approach has been made to work [Chu84], it has proven so clumsy that it is no longer considered viable. Practical technology updatable generators all use the compacter, but do so in different ways. Each way has its advantages and limitations, and in general, several different ways might be used in the construction of a chip. The methods fall into three categories.

In the first one the generator author uses the editor to create one or more symbols on the virtual grid, compacts them onto the fixed grid, and writes them out as IMAGES language symbols. These symbols are then used like any other set of fixed-grid IMAGES symbols, and reside as text inside an IMAGES-C generator source file. Each time the generator is invoked the IMAGES-C language, with or without the router, is used to glue together instances of the symbols. One disadvantage is that this method does not provide any automatic parameterization of the cell pitches or of the transistor sizes.

Under this discipline it is necessary for the generator author to recompact the symbols each time the technology changes but not each time the generator is run. To simplify the administration of the CAD system the compacter can be run once for each technology change as controlled by the UNIX utility "make." ** This mechanism would be most appealing if the compacter were extremely slow running (as might be the case for a tight two-dimensional compacter). But since most compacters are reasonably fast its utility is mostly limited to the case where there are some hand designed symbols in a design, such as a memory cell or sense amplifiers in a memory. In general these have special requirements and cannot be compacted. Each time the technology changes such a generator must be "touched up": the compacter can recompact the perimeter and control cells to match the new pitch of the array, but some of the core cells must be revalidated (at least). This method is illustrated in Figure 5-5, and the output in 5-6.

---

** Make checks for the most recent date of change on a file and runs a specified program that is normally a compiler, but in this case would be the compacter.

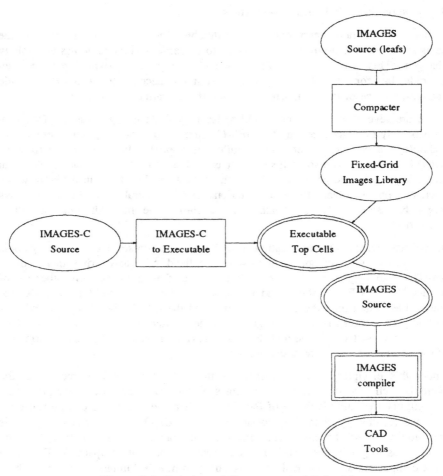

Figure 5-5: Relation of Compacter in Method 1.

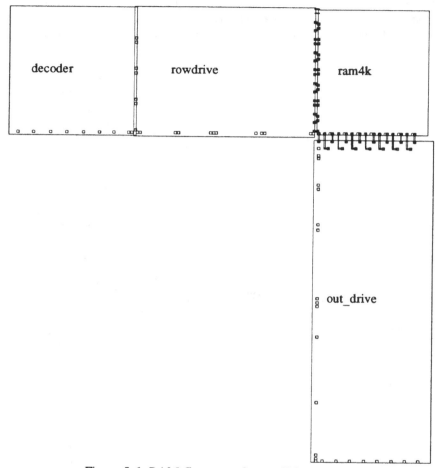

Figure 5-6: RAM Generator Output Using Method 1.

The second mechanism uses the compacter each time the generator is invoked. The author creates some symbols in virtual-grid IMAGES. Each time the generator is invoked they are compacted (perhaps with pitch matching). They are then wired together using the IMAGES-C language/router in the fixed grid.

This method is based on the idea that some design rules are not likely to change much beyond simple, linear scaling (a.k.a. "lambda shrink"). It is therefore possible to simplify the generator process by including some features such as routing explicitly in the IMAGES-C language. Of course, the "glue" wiring that is described in IMAGES-C code is not guaranteed to be technology updatable.

However, when spacings are parameterized by a distance specified in the technology file, if a technology change should make this layout illegal, the IMAGES-C file would merely need to be recompiled. This method is illustrated in Figure 5-7.

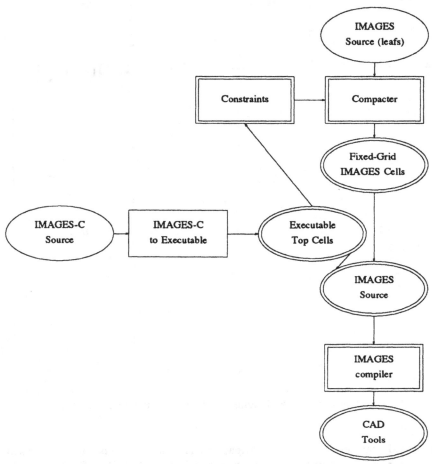

Figure 5-7: Relation of Compacter in Method 2.

One example of this method is an up-down counter generator parameterized by the number of bits and its drive capability. The size of the output drivers can be specified to the program based on the amount of capacitance the buffers will drive. The layout for a 2-bit counter with small output drivers is shown in Figure 5-8.

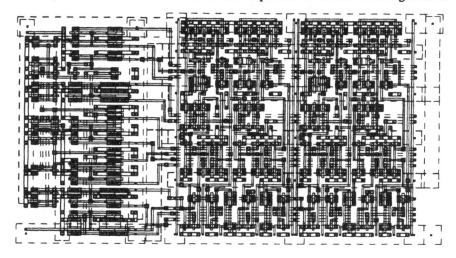

Figure 5-8: Counter Generator Built using Method 2.

The third method of writing a generator depends even more on the compacter. The author writes IMAGES-C code that compiles into one, or several, virtual-grid IMAGES files. Each time the generator is invoked, all the IMAGES files are compacted and pitch matched.

This mechanism is the most costly in terms of execution speed because the compaction and pitch matching operations must be performed for the entire symbol at each execution. Its advantage is that it guarantees a layout free from design rule violations, and can provide extra parameterization at almost no cost. In particular, because the design passes through the compacter, the user can program the IMAGES-C file to expand the pitch between input or output pins, thus allowing the circuit to match with symbols produced by other means (*e.g.* by hand, or by non-compacted generators). The designer can also depend on the compacter to adjust the symbol for varying transistor sizes, since these do not affect the virtual-grid topology. This method is diagramed in Figure 5-9, and is the one used by the SC2D system discussed in the next chapter.

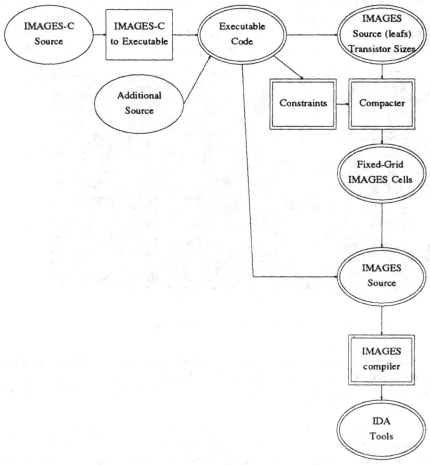

Figure 5-9: Relation of Compacter in Method 3.

## 5.5 Summary: Advantages and Limitations

Designers who work with generators have traditionally recognized some frustrating limitations. One is the difficulty of getting the generator to produce anything useful, since, as is illustrated in Figure 5-10, there are so many steps between the source accessed by the author and the final output. Perhaps more seriously is the limitations encounted even after the generator is finished. Even though a generator can take several parameters, there is often some critical aspect of it that is not parameterized. This limitation may make it unsuitable for a particular application (*e.g.* the input and output wires are on opposite sides, when they really need to be interleaved, or *vice versa*). This problem is especially acute for low-level generators, such as registers.

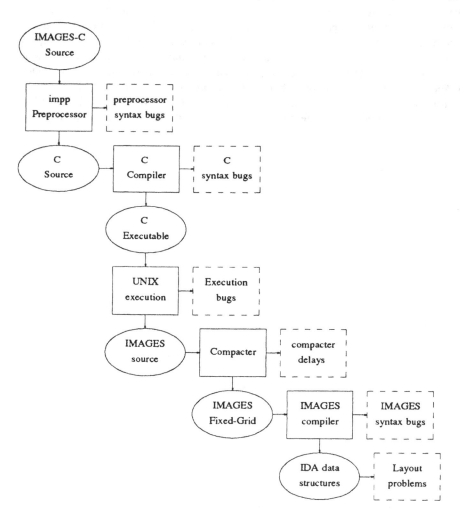

Figure 5-10: Potential Problems in Writing and Using A Generator.

One way to approach this is to have the largest library of generators possible. The purpose of the IMAGES-C extension is to expedite the creation of new generators and support their porting to new environments, particularly to new technologies.

But this does not solve the problem that designers who write generators are still placing much effort and intelligence in solving a single problem, and not in actually building the chips that "pay the rent." In practical terms, it is important not to tie up the best designers writing, modifying and maintaining generators. So the current thinking is to limit the amount of work on generators by limiting the number of supported generators as much as possible. In fact, only four are in common use: PLA's, ROM, RAM, and the pad frame generator. Just keeping these up-to-date requires a substantial ongoing effort. (Other generators reside within individual designer's libraries, but without detailed documentation needed to keep them available to the public.) In contrast, the logic-based generators can be used over a much broader class of problems. That is why so much effort has gone into them over the last few years, and that is the subject of the next chapter.

# Chapter 6
# Automatic Layout of
# Switch-Level Designs

Most VLSI layout styles in use today fall into a few broad categories. These include dedicated site styles such as gate arrays and sea-of-gates, prepared library styles such as standard cells, gate-matrix [Lope80], and full-custom design. Each of these has advantages and disadvantages in terms of performance versus design time, with gate arrays generally being the quickest to obtain and full-custom design being the highest performance. The Silicon Converter system is a tool designed to produce compact and electrically high-performing MOS layouts. It operates somewhere in the range between standard cells and traditional full-custom design. Like standard cells, it can produce a layout with almost no designer intervention. Like traditional full-custom design styles, it can handle arbitrary transistor-level circuit specifications and is not limited to those found in a library. It can take advantage of regular structures if present, or dedicated floorplans if they are available. Naturally, the layout for a given function may not be as tight as the best special purpose generator (*e.g.* PLA, ROM, RAM, *etc.*), but it can be highly effective for a wide range of design situations.

## 6.1 System Organization

To use the silicon converter system the designer must supply a complete switch-level specification of his/her design. (In this chapter, the words *switch-level* and *transistor-level* will be used interchangeably.) The medium for this specification is the "simnet" language used by many of the other tools. Once the designer has produced a transistor-level representation of his or her design, it can be validated with SOISIM and sized with TILOS, as discussed in Chapter 7. SC2D then places and routes the symbol, splitting the design into N rows, and uses the strip layout program "SC2" to lay out each row in detail. (N may be specified manually, or may be chosen by SC2D to fit a requested aspect ratio.) SC2D also produces a geometric floorplan in the IMAGES language that unites the layouts created by each SC2 run. This system organization is shown in Figure 6-1.

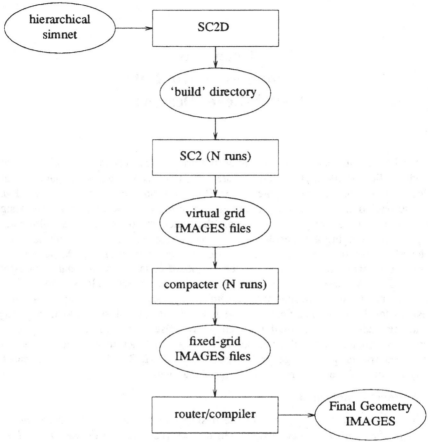

Figure 6-1: System Architecture of SC2D.

**Where Do You Get the Switch-Level Models?**

Since the silicon converter system accepts any circuit, one way to use it is to take an existing layout, extract it to generate a switch-level model, and synthesize it. This approach has the advantage that existing designs can retain the correctness of their circuit model, while taking advantage of electrical and/or layout improvements created by TILOS and SC2D. This is most likely to occur when new geometric requirements are applied to a working part of a chip.

On the other hand, new designs can be entered with a schematic capture system that operates with a library of predefined cells such as a standard cell library. This method does not take full advantage of the extra degrees of freedom possible with a switch-level layout tool, but can still make use of the TILOS transistor sizer. Novel circuits not found in the cell library can be entered with transistor-level schematics, but this tends to be tedious.

A third technique makes use of the Switch-Level Integrated Circuit Compiler (SLICC). SLICC supports the specification and simulation of arbitrary circuits at the switch level. Newer designs need never be represented as standard cells, but can be specified from the beginning in "C" language syntax. They can therefore include logic cells not found in any standard cell library. SLICC is discussed in Chapter 7.

## 6.2  Steps in SC2D

Once the circuit design is available, the layout processing begins when SC2D parses its input file(s) and builds data structures that reflect the source-level logical hierarchy. SC2D can use these structures to guide placement at the global stage. Later, after the design has been broken into transistor pairs, the storage associated with the original hierarchy is reclaimed.

The most critical part of layout is placement. In SC2D several methods of placement are available. The first is manual: If the user specifies a placement file it is used to position the logic gates mentioned in it. Such a file contains lines identifying a logic gate by name, a row number and a relative horizontal position within the row. The row number is absolute, but the horizontal position number may be adjusted by SC2D in the face of missing or conflicting information. This adjustment is done using an initial placement heuristic. The placement file is normally used when the designer has a floorplan clearly in mind and wants complete control over the major data and control flow. (*I.e.* data shall run horizontally, and control shall run vertically.) If no placement was specified SC2D performs its own placement making use of up to three sets of heuristics. After each heuristic a cost is estimated and the winning position is recorded.

### Automatic Placement - First Estimates

The first placement algorithm is a "level 0" heuristic: The incoming cells are simply snaked around from one row to the next in the order in which they were expanded from the input logic hierarchy. This is done to get a first cut at the cost of the layout. In SC2D a placement's cost is figured as the sum of all *net costs*. A net cost is the half perimeter of the smallest rectangle enclosing all points on a net. This represents an estimate of the routing costs on a net-by-net basis. When net costs are calculated for an externally accessible net, the enclosing rectangle is expanded to touch the side(s) where access was requested. Because objects that are close logically tend to be close textually, and therefore end up being close geometrically, the snaking technique works surprisingly well in some designs. Since its computation cost is trivial, it is always performed prior to other heuristics. After this comes a slightly refined variant: the cells are put in a linear order using min-cut partitioning, and the resulting logic is again placed using the snaking method. This is evaluated and compared with the first placement, and the one with the higher cost is discarded.

## Making Use of Logic Hierarchy

The second major category of placement algorithms in SC2D are those that make use of logic hierarchy expressed in the original language. SC2D trys a series of mappings that map the logical space, in one, two or more dimensions into a floorplan. The dimensionality of the logic space is the depth of the logic. This mapping is done by assigning to each logic cell a vector of non-negative integers corresponding to its position in the logic hierarchy. For example, consider a 4X4 combinational multiplier. This can be specified hierarchically as four word-circuits, each consisting of four one-bit circuits, where each bit circuit consists of a full adder enhanced with other circuitry. If this specification were read by SC2D, each adder would be assigned a vector of length two. The first number would indicate the word it was in and the second would indicate the number of the bit within the word, from most significant to least significant. SC2D would then attempt to map these 16 cells onto the required number of rows, while preserving some type of symmetry if possible. To do this SC2D uses Algorithm 6-1.

```
/* accept an array of cells with logical positions attached,
** and place them */
place_regular_cells(cells)
    /*  E is a vector of integer weights, with length
    *** equal to depth of logic*/
    lowest_cost = ∞
    for Weights in {P | P is a permutation of E}
        for i in 0 .. num_cells
            /* each cell has a vector Logic_pos associated with it,
            ** and a field "prod" that stores the dot product */
            cells_i.∏ = dot_product(Weights, cells_i.Logic_pos)
        end for

        sorted = sort(cells) /*based on the "prod" field */
        /* span is a factor of the number of cells */
        for span in {x | Nx = num_cells}
            for i in 0 .. num_cells
                cells_i.position = assign_pos(i, sorted, span)
            end for
            this_cost = estimate_layout_cost(cells)
            if this_cost < lowest_cost  then
                remember_layout(cells)
                lowest_cost = this_cost
            end if
        end for

    end for
end place_regular_cells
```

Algorithm 6-1: Regular Design Placement.


This technique requires exponentially greater time with higher dimensions of logic depth, since the set of possible of mapping vectors and spanning values grows as the factorial of the length of such vectors. However, it has not proven to be a problem for practical designs since regular structure is most often found with order one or two, only occasionally with order three and rarely with four, which is the upper limit to SC2D.

## Placement of Unstructured Logic

The structured placement technique outlined above seems to work best when the layout is completely regular, without random logic cells breaking symmetry, and the number of rows requested divides the total number of cells exactly. Such is the case in some arithmetic and logical units, but seldom in control units. For these, SC2D uses a min-cut technique [Dunl83] for initial placement, followed by an optional improvement step using simulated annealing. The simulated annealing schedule may be controlled by the designer in four ways:

1. Specifying a starting temperature.

2. Specifying a number of moves per cell.

3. Specifying the ratio of one temperature to the next.

4. Setting a specific limit on the number of CPU seconds to spend on it.

The default is to run min-cut first, with itself using a few random starts, followed by a small clean-up period of simulated annealing corresponding to a low temperature "quench."

## Global Routing

Following placement of the logic cells, SC2D breaks each cell into pairs of devices (one pFET paired with one nFET), splits the wide devices, and does inter-row routing. In the row-based layout structure produced by the system, global routing consists largely of adding feedthrus at appropriate points, and deciding which track(s) a given signal should run in, in those cases where there is more than one candidate. This feedthru insertion is done by calculating the vertical range of each net (from the lowest row that it appears on to the highest), and by inserting feedthrus on intervening rows when necessary. (If a signal appears on complementary pFET and nFET gates, an additional feedthru is not required). The choice of where to place the feedthru within a row is controlled by a series of heuristics. These heuristics make use of the horizontal intervals of each net on each row. (The *interval* of a net is the region between the leftmost and rightmost points on that net.) An attempt is made to minimize the routing overhead of each net by placing the feedthrus inside the intersection of the horizontal intervals from adjacent rows if this intersection is not empty. Furthermore, wherever possible the feedthrus are placed directly above each other, so that no routing track will be required. Finally, when feedthrus are being positioned a check is made to see if they disrupt a diffusion sharing between adjacent devices, which would increase the row's width. When this happens, SC2D searches the neighborhood for a position that is a natural break in the diffusion.

## Intra-Row Routing Improvement

In implementing horizontal connections, SC2D has two alternatives: metal wires can be run in the internal region between the pFET and nFET pairs or in the region between rows of logic. SC2D makes the decision of which to use based on the spacing of points on the current net within each row. For example, a cluster of three points on a net that are close to each other on the left end of a row would be routed with a wire in the pFET/nFET gap. On the other hand, two such clusters located at opposite ends of a row would be routed with a wire in the inter-row routing region.

SC2D generates this routing by considering each net separately. If the net under consideration is local to the row the clustering distance is set to one value (such as 8), and if it used externally to the row it is set to a smaller distance (such as 5). Distance is measured as a number of intervening transistor pairs. Each row is scanned from left to right and the points on each net are grouped into clusters when they are less than the clustering distance apart. Other factors that govern the heuristic include:

1. Whether the cluster has a transistor gate (or feedthru) associated with it. (Clusters without access to polysilicon gates require more complex structures to provide external connections.)

2. Whether the transistor pairs involved in the cluster are balanced (having both a pFET and an nFET) or unbalanced.

3. Whether the row being processed is even numbered or odd numbered.

The last consideration is necessary because odd-numbered rows are flipped about the $x$ axis to minimize the space wasted by tub polarity conflicts, and the intra-row routing added by this algorithm always goes above an odd numbered row (the bottom row is numbered 0). Extra feedthrus are added, if necessary, to allow connections among clusters.

The heuristics for moving wires out of the pFET/nFET gap and into the routing region can have a significant effect on the compactness and performance of the resulting layout. To illustrate this, consider what would happen without it. If all connections were done internally to the row, the pFET/nFET gap would be increased by several routing tracks. If all connections were done externally, the compactness of the layout would be degraded with poly wires leading in and out of the layout strips to connect nearby cells, as happens in standard cells. These poly wires can also cause unnecessary breaks in the diffusion. But if the job is done correctly it tends to decrease the height of the layout because the inter-row routing region can be shared by more nets. This increases the probability that short wires can fit in the gaps left by others (*i.e.* there is an economy of scale in routing regions) and the likelihood that a horizontal track can carry the same net for both the row above it and the row below it. It may also improve performance because it reduces the length of all poly wires that bridge the gap between the pFETs and nFETs within a row. A wide gap increases the capacitance and resistance of hundreds of nets simultaneously, affecting even the most compact and local nets with the smallest drive and load transistors. The effect of this heuristic can be seen in Figure 6-2, where three tracks are saved by sharing them.

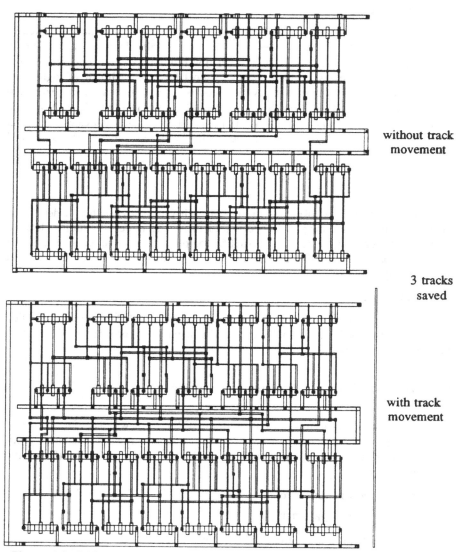

without track
movement

3 tracks
saved

with track
movement

Figure 6-2: SC2D Layout Comparing Area with and without Intra-row Routing.
The bar on the right side of the lower layout is
the same height as the top layout.

## 6.3 Steps After SC2D

SC2D builds a set of files in a dedicated subdirectory that is automatically created when SC2D is run. These files include an overall floorplan (specified in IMAGES) which contains the detailed inter-row routing specifications, but not the routing wires themselves. It also contains simnet files for each row of logic, plus a UNIX "makefile," a journal file (which reports warning messages and execution time), a final position file (which can be reused on the next SC2D run), and an estimated cost file, which contains a single number that indicates the final estimated cost of the layout. To proceed with the design process, the user need only type

```
make all
```

This causes UNIX to invoke SC2 to lay out each row of transistors in detail, the compacter to bind the current technology to them, the routers to supply detailed routing between rows, and the IMAGES compiler to resolve geometrical and electrical constraints in the overall layout. The final result is a set of fixed geometry layout files.

## 6.4 Detailed Layout in SC2

When SC2 is used under SC2D it makes use of the detailed placement specified by SC2D. But SC2 can also be used as a stand-alone program with its own placement capabilities.

### SC2 Transistor Placement, Splitting and Flipping

In the absence of a specific placement from SC2D, SC2 will lay out a strip of logic using local heuristics. Unlike SC2D, SC2 requires a flattened (non-hierarchical) description of the transistors to be laid out. It then performs the following steps:

1.  It splits the wide transistors.

2.  It pairs pFETs with nFETs.

3.  It places them in order from left-to-right,

4.  It flips them (source/drain) to minimize diffusion mismatches.

5.  It routes among them.

6.  It writes out a virtual-grid IMAGES file.

## Splitting

The transistor splitting mechanism breaks wide transistors into several small parallel transistors. To do this, it must decide on the maximum width to be allowed before splitting. SC2 uses a search technique that tests all the reasonable splitting points, and determines an estimated total area cost for each (Algorithm 6-2). The area is the product of the estimated width and height. The smaller the maximum width for transistors, the shorter the height will be. But at the same time, the width will be greater because there are more columns. NFETs are processed first, simply because they tend to be larger and have a greater effect on the height. Because of splitting, they also tend to be the limiting factor in the width of the cell.

```
/* accept proposed width limits, return estimated area */
area(pfet_limit, nfet_limit)
   p_columns = ∑ effective parallel pFETs
   n_columns = ∑ effective parallel nFETs
   if p_columns > n_columns then
      w = p_columns
   else
      w = n_columns
   end if
   h = pfet_limit + nfet_limit + routing_est
   return wh
end area
```

```
/* accept a set of pFET and nFET widths, and determine
*** the best limits for transistor widths, nfet_lim and pfet_lim */
split(pfets, nfets)
   est_n = (∑ nfet.width) / num_nfets
   min_a = ∞
   for l in { w | w evenly divides a pFET width}
      a = area(l, est_n)
      if a < min_a then
         pfet_lim = l
         min_a = a
      end if
   end for

   min_a = ∞
   for l in {w | w evenly divides an nFET width}
      a = area(pfet_lim, l)
      if a < min_a then
         nfet_lim = l
         min_a = a
      end if
   end for
end split
```

Algorithm 6-2: SC2's Transistor Splitting.

## Flipping

The next step is to orient, or "flip," transistor pairs to maximize the number of adjacent sources and drains sharing the same signal. SC2 makes use of a dynamic programming algorithm to do this. It starts on the left end and moves right, while keeping track of all configurations that have a chance of being optimal, and discarding all others. Although the set of possible flippings grows exponentially in the number of pairs, the breadth-first search can be pruned tightly, which results in a run time that grows linearly. The pruning of the search tree is based on the observation that as far as the remainder of the layout is concerned, a configuration of the first M transistors is completely characterized by its cost and by its rightmost net, which is the only exposed connection. No subsequent steps can ever justify a more costly configuration with the same exposed net. Since there are only two possible exposed nets (resulting from the current transistor being flipped or not flipped), there are only two configurations worth considering.

Therefore, as SC2 works from left to right it considers only the "current" transistor, and views it in light of the two configurations produced by the preceding step in the algorithm. The preceding step produced two configurations, with two different nets visible. SC2 takes these two and generates four configurations. These are the two original configurations with the current transistor added on in both the flipped and non-flipped orientation. Of the two configurations with the current transistor flipped, SC2 prunes off the more costly. Likewise, of the two configurations with the current transistor not flipped, SC2 prunes off the more costly. These two configurations represent the optimal ways to orient the transistors and allow access to the two different nets on the right end. SC2 then steps right and repeats the process. When it reaches the right side of the symbol it has two configurations, and it can choose the minimum cost option, which is guaranteed to result in the fewest source-drain mismatches globally. This is shown in Algorithm 6-3.

When two configurations produce the same number of diffusion breaks, SC2 uses secondary heuristics to select one or the other. For example, if two pFETs share a signal and power, then SC2 will flip the transistors to share the signal node. This flipping reduces the amount of capacitance on the shared node and adds to the capacitance of the Vdd node.

```
/* cost(C) returns the number of diffusion breaks in configuration C*/
cost(configuration C)
    breaks = 0;
    for i in 1 .. length_of(C)
        if C_i.drain ≠ C_(i-1).source then
            breaks++
        end if
    end for
    return breaks
end cost

progress(i, configuration C 1, C 2)
    if i = num_of_fets then
        if cost(C 1) < cost(C 1)  then
            accept(C 1)
        else
            accept(C 1)
        end if
        DONE
    end if

    TMPC 1 = append(C 1, "non_flipped")
    TMPC 2 = append(C 2, "non_flipped")
    if cost(TMPC 1) < cost(TMPC 2) then
        WC 1 = TMPC 1;
    else
        WC 1 = TMPC 2;
    end if
    TMPC 1 = append(C 1, "flipped")
    TMPC 2 = append(C 2, "flipped")
    if cost(TMPC 1) < cost(TMPC 2) then
        WC 2 = TMPC 1;
    else
        WC 2 = TMPC 2;
    end if
    progress(i+1, WC 1, WC 2) /* move to the right */
end progress
```

Algorithm 6-3: Algorithm for Ordering FETs to Minimize Breaks.
(assume that "configuration" variables ($C$ 1,$C$ 2,$TMPC$ 1,$TMPC$ 2,$WC$ 1,$WC$ 2)
are lists of booleans representing "flipped" or "non_flipped" states,
and invoke the system by calling "progress(0,nil, nil)")

## Intra-Row Routing

Once the position and orientation of transistors is complete SC2 does the detailed internal routing. This is accomplished with a special router that incorporates heuristics tuned to intra-cell routing. One requirement of this router is that it must always complete 100% of the nets. The router can use extra area if necessary, but since it is supporting several higher levels of CAD tools it must not produce any shorts or missing connections.

To do this SC2 uses a "greedy" router that operates orthogonally to the way most greedy routers work. Instead of running several nets in parallel from left to right, it runs one net at a time to completion, and places the tracks from bottom to top of the routing region. It starts by ordering the nets such that those most closely connected to the nFETs are routed first, followed by those with more connections to the nFETs then the pFETs, followed by those with more or the same number of connections to the pFETs, and finally those just connected to the pFETs. Within each of these four categories the nets are ordered by length, routing the shortest nets first. This enables the router to use the tracks closest to the FETs for short connections on the same side.

The router operates on a special coordinate system where the routing area can expand arbitrarily without effecting the positioning of the transistors. This avoids the overhead (found in some other systems) of interrupting routing and moving cells when the router runs out of space. The SC2 router does not consider order constraints (a.k.a. *cycles*) when routing each net. Therefore, it can produce shorts where two nets intersect each other on the same column (Figure 6-3a). These are detected in a post-route pass and corrected by moving the connection point on the pMOS side (Figure 6-3b). Because of the initial ordering of nets to be routed (nMOS side first from the bottom, pMOS side last at the top) this situation does not occur very frequently, and the overhead entailed is small. In fact, when a constraint-based compacter is used, most of the extra width is recovered by merging it into adjacent routing.

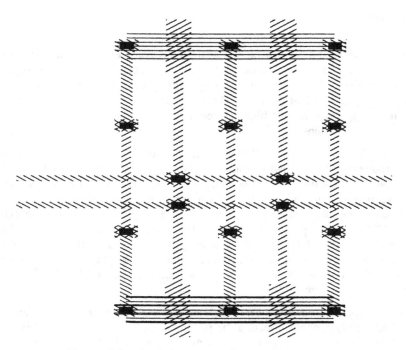

Figure 6-3a: Cycle Introduced by the SC2 Router.

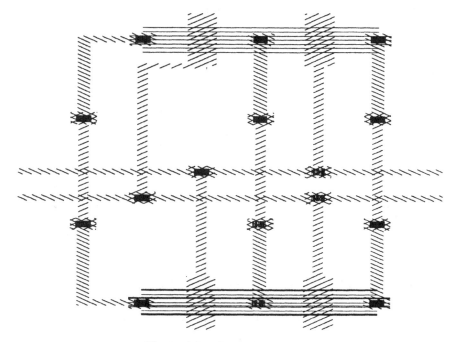

Figure 6-3b: Cycle Corrected.

## Detailed Routing with Second-Level Metal

When the original Gate-Matrix style was invented, only one level of metal was available. Today, most processing lines support two or more levels. Several systems have attempted to expand on the original style to make use of the second level [Gee87]. The need for this is generally not motivated just by area savings, but by performance, since second-level metal has lower resistance and capacitance than polysilicon.

The silicon converter system has experimented with several such styles. The first style of layout developed substituted metal2 for poly directly wherever this was possible (Figure 6-4). This made vertical connections within the cell, feedthrus, and external connections on the top and bottom, in metal2.

The second style of layout uses metal2 horizontally and poly and metal vertically (Figure 6-5). Power is now distributed on metal2. This scheme allows for the diffusion areas to drive outputs directly in metal. The only use of poly is for connecting the gates. The net result of either scheme is an all-metal path from the source or drain of the largest transistors all the way to the gates of the driven transistors. Such a low resistance path is essential to realizing the full benefits of TILOS transistor sizing.

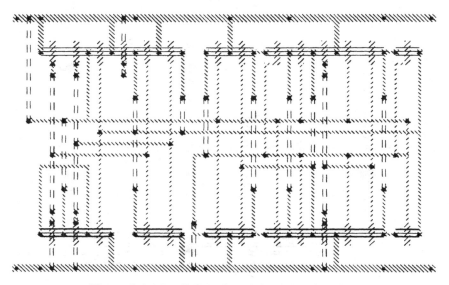

Figure 6-4: Metal2 Substituted for Poly Directly.

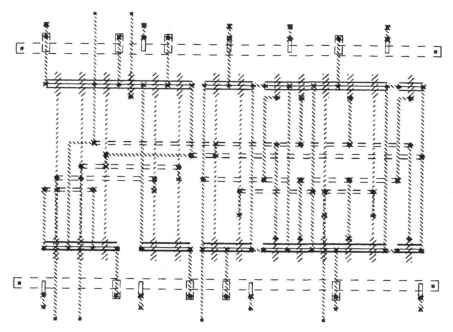

Figure 6-5: Metal2 Used Horizontally.

## Via Elimination

As a post-routing process SC2 touches up and improves the routes through wire merging and via elimination. To do these geometric operations the indirect coordinates used by the router are first converted into numerical positions on a virtual grid. Each object (wire or contact) is then converted into a set of left and right edges and the set of edges is sorted. A scan-line (similar to the one described in Chapter 4) then passes from left to right, and as it sees the left and right edges of objects it calculates the interaction of objects at that $x$ coordinate.

The first operation is to merge any redundant or overlapping wire segments. The next phase does the opposite, breaking the merged wires into smaller segments connecting perpendicular wires and contacts. The merging and breaking process guarantees that exactly one wire will exist between any two points that need to be connected, and simplifies the via elimination process. Once the wires have been broken into segments a linked list is created that records the objects intersecting a given wire or contact. This list is the adjacency graph used in classical via elimination.

The elimination of vias operates under a set of heuristics tuned to MOS technology. The process is greedy and monotonic: if a wire changes levels it cannot change again. The via eliminator loops in the following cycle until no wires change level:

1.  change poly wires into metal wires (always)

2.  change metal wires into poly wires (if they are short and this eliminates a via)

3.  change metal2 wires into metal wires (if this eliminates a via)

4.  change metal wires into metal2 wires (if this eliminates a via)

Algorithm 6-4: Simple Via Elimination Heuristics.

A wire can change level if the new level does not conflict with any of the overlapping wires. A via can be eliminated if all the wires that intersect it are of the same level (Figure 6-6). Situations occur where a via cannot be eliminated by simply changing the a single wire. This is where the adjacency graph information is used, because it allows the software to examine a wider context efficiently.

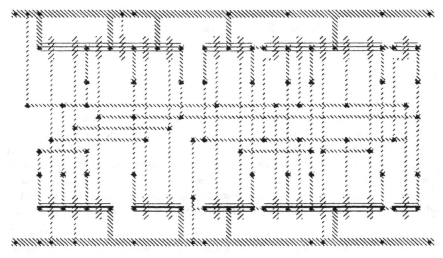

Figure 6-6a: Routing Before Via Elimination.

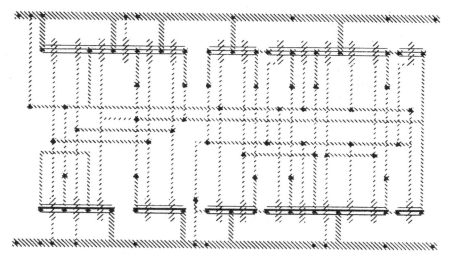

Figure 6-6b: Routing After Via Elimination.

As a final cleanup step the wires are once again merged where this is possible, with one exception. Vertical wires that intersect with, but do not end on a horizontal wire become two vertical segments. The reason for this is that the IMAGES output may need to be extracted by the IDA virtual-grid circuit extractor. As discussed in Chapter 4, if the wires were allowed to cross, they would be extracted as two distinct nets, and the compacter would refuse to compact it.

## 6.5 Performance and Parallel Processing

For jobs where placement is not a major problem SC2D normally completes in a time that is much less than subsequent processing stages. On the other hand, if high-temperature simulated annealing is used for placement it can consume an almost unlimited amount of computing resources. To contain this problem SC2D includes two features. The first is based on the observation that at high temperatures simulated annealing usually just succeeds in scrambling the design: real gains occur only after the temperature starts to cool. Therefore SC2D provides a quick scrambling algorithm based on a pseudo-random number generator and a user-supplied seed: the simulated annealing algorithm itself is not used for this purpose. Secondly, SC2D is supported on a multi-processor configuration consisting of several workstations connected via Ethernet. Since the placement part of the run tends to be CPU intensive, this loosely coupled arrangement seems to be reasonably effective even with the limited bandwidth available between processors and between processors and the common file server.

A designer can make use of these computing resources by starting several independent SC2D runs in parallel, where each has different parameters such as the number of rows to use, simulating annealing schedules, or random number seeds. The best of these runs, as indicated by the estimated cost file mentioned previously, is then implemented using the same collection of workstations running in parallel. This time each workstation does the detailed layout of a single row using SC2 and the compacter. While this approach does not speed up any individual run of SC2D, it does allow a wider design space to be searched more easily. Furthermore, it speeds up the subsequent processing, which typically represents more CPU time, in aggregate, than the SC2D run, especially in the compacter phase. Under the Network File System (NFS), all the processors appear to be working under a single UNIX directory. Their actions are coordinated under a software system that automatically queues jobs and hands them to workstations as they become available. This method is diagramed in Figure 6-7.

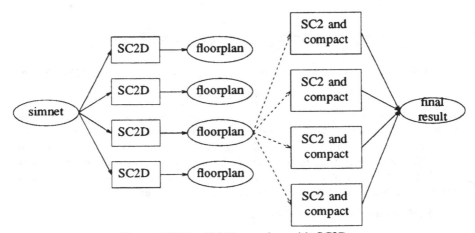

Figure 6-7: Parallel Processing with SC2D.

## 6.6 Summary

SC2D has successfully laid out silicon for designs ranging from a few dozen transistors up to $\approx$ 40,000 devices. In general the layout quality seems to be dominated by placement: a poor placement produces a poor result; a good, and "natural" placement produces a result comparable to hand design.

The efficiency of the intra-row/inter-row routing heuristics and SC2 seems to compare satisfactorily with the average efficiency of hand design. In one experiment done at Bell Labs, SC2D produced a smaller layout than hand design for a multiply-by-constant module. This was because the hand design left gaps where the null circuitry (corresponding to multiplying by 0) would have been, while SC2D automatically moved other circuits to fill them in. But the biggest advantage seems to come when SC2D is used with transistor-level optimization tools such as TILOS that produce highly tuned circuits with little or no regularity in them. While one would think of special control circuitry as falling into this category, even arithmetic circuitry can make use of it too. For example, for maximum speed a 32-bit adder can be generated where each bit is slightly different because of timing considerations. Such circuits are tedious and costly to lay out manually, and would probably be impractical without a tool such as TILOS and the silicon converter system.

# Chapter 7
# Switch-Level Tools

The tools and techniques discussed in preceding chapters have addressed the issue of turning a circuit into geometry. But higher level issues remain: How does one get the circuit level design in the first place? How does one tune it to reduce power and improve performance? And, how can a designer take advantage of the unlimited variety of circuits possible with custom VLSI? This chapter addresses these issues.

## 7.1 Transistor Sizing with TILOS

Full custom VLSI offers the opportunity to tune every transistor to optimize its performance. Unfortunately, the problems of tuning even a medium size circuit by hand can become overwhelming. One problem is that increasing the size of one transistor to increase its drive capability has the unwanted side effect of adding capacitance on its gate, thus slowing down its input. Thus designers working by hand typically use a few simple rules of thumb (*e.g.* make drivers in a chain exponentially larger) or spend many hours of human and machine time iteratively tuning and reevaluating.

However, the problem is not as intractable as it may seem. In fact, it can be formalized fairly simply. Given a synchronous CMOS circuit of the form shown in Figure 7-1 with N transistors of sizes (channel widths) $x_1$, $x_2$, ..., $x_N$, the following question is considered: How can the circuit's performance be improved by adjusting the $x_i$? Two figures of merit are of special interest, the time "T" and the total transistor active area "A." T is defined to be the minimum clock period at which the circuit will operate correctly. T is therefore the maximum, over all possible flip-flop to flip-flop paths, of the signal propagation delay along the path. The other quantity, A, is simply the sum of transistor sizes. A is positively correlated with a number of other attributes of the circuit that should be minimized or constrained: These include silicon area, capacitance-discharge power, short-circuit power (power dissipated due to current flow from Vdd to Vss during the short period of time when both the pullup and pulldown networks are partially turned one) and probability of a device failure within a chip.

TILOS (TImed LOgic Synthesizer, pronounced *tee-los*) is a program that accepts a transistor connectivity file and I/O-delay file, and adjusts transistor sizes and connectivity within logic gates to meet the user's requirements for T while minimizing A. TILOS's output is a transistor connectivity file, with the new transistor sizes, that can be passed to a layout program such as SC2D.

TILOS contains a static timing analyzer which recognizes latches. It is thus capable of extracting all relevant timing paths from a circuit of the form shown in Figure 7-1. This recognition process is similar to that used in other static timing analyzers ([Agra82], [Oust84], [Joup84]) and will not be further described here.

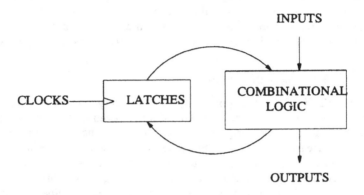

Figure 7-1: Memory/Combinational-Logic Model of Digital CMOS Circuits.

Several authors ([Rueh77], [Glas84], [Hedl84], [Lee84], [Mats85]) have reported on optimization techniques for transistor sizing. The advantages of TILOS over previous approaches include:

1. Using a distributed RC model of delay, signal delay along a logic path is shown to be a *posynomial* function of the transistor sizes. (Roughly speaking, a posynomial is a multivariate polynomial with real exponents and positive coefficients. For details, see [Ecke80].) As will be explained shortly, a remarkable consequence of this fact is that when the transistor sizing problem is viewed as a constrained optimization program, a local optimum must necessarily be a global optimum. Path delay remains posynomial if wire widths are also considered to be variables.

2. TILOS couples static timing analysis with transistor sizing, relieving the user of the need to specify which paths are to be optimized. Rather, the user specifies desired behavior of I/O signals, including clocks, and TILOS determines what paths are in need of improvement.

3. Transistors in latches are sized, as well as transistors in combinational gates.

4. TILOS sorts series-connected subnetworks in each complex gate so that the subnetworks with earlier-arriving inputs are closer to the power supply. This heuristic gives transistor sizing a chance to operate: A transistor with an earlier-arriving input can be made larger and still have time to turn on before other inputs arrive, providing a lower-resistance path to power. In addition, the increased source and drain capacitance of the larger transistor helps, rather than hinders, the output transition.

## MOSFET Model

The MOSFET model that TILOS uses is shown in Figure 7-2. The gate, source, and drain capacitances are all proportional to the transistor size X, and the source-to-drain resistance is inversely proportional to X.

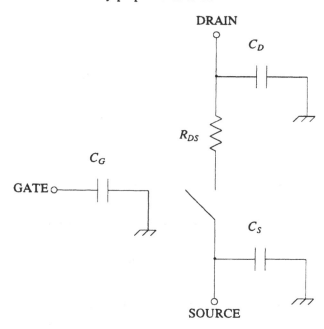

Figure 7-2: TILOS's Electrical Model of a Transistor.

## RC Delay Model

Figure 7-3 illustrates the modeling of gate delay by a distributed RC network [Penf81, Lin84]. In the RC network shown, an upper bound for the discharge time is

$$(R_1+R_2)C_2+(R_1+R_2+R_3)C_3. \tag{1}$$

This represents a much tighter bound than a lumped R and C model.

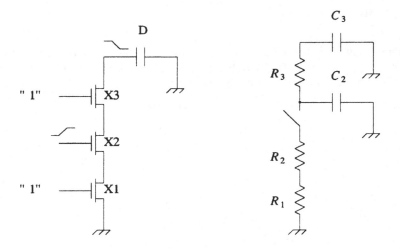

Figure 7-3: Delay Through Pulldown Network Modeled with RC Network.

## The Transistor Sizing Problem Formulated as an Optimization Program

Let K be a positive constant, and let A and T be as defined above. Consider the following optimization program:

Minimize A subject to the constraint T&lt;K.

This formulation can be used when the circuit must fit inside a system with a given clock period K.

With the MOSFET and RC-delay models, the delay through any series configuration of transistors can be expressed in terms of the transistor sizes and wire capacitance. The important point here is the form of (1) when expressed as a function of the transistor sizes $x_i$: Each $R_i$ is proportional to $1/x_i$, and each $C_i$ is some constant (for wire capacitance) plus one term for each transistor whose gate, drain or source is connected to the node. This term is proportional to the transistor size. Thus (1) can be rewritten as:

$$(A/x_1 + A/x_2)(Bx_2 + Cx_3 + D) + (A/x_1 + A/x_2 + A/x_3)(Bx_3 + E) \qquad (2)$$

where A, B, and C are constant coefficients for resistance, drain and source capacitance, respectively, and D and E are wire capacitances. It is interesting to note that if wire widths as well as transistor widths are treated as variables, then the expression for delay remains in the same form as (2). This means that the widths of wires could be optimized at the same time, allowing wider wires only where needed.

**Delay Through a Complex Gate**

For each transistor in a pulldown or pullup network of a complex gate the greatest resistance path from the drain to the gate output is computed, as well as the greatest resistance path from the source to a supply rail. Thus for each transistor the network is transformed into an equivalent series configuration, and the calculation of the previous section is applied.

**Delay Through a Single Circuit Path is a Posynomial**

Associated with every circuit path are two path delays: one for the case where the input to the path is rising, the other for the falling input. Since a path delay is simply a sum of gate delays as in (2), the general form of a path delay is as follows:

$$\sum_{1 \le i,j \le N} a_{ij} \frac{x_i}{x_j} + \sum_{1 \le i \le N} \frac{b_i}{x_i}, \qquad (3)$$

where the $a_{ij}$ and $b_i$ are nonnegative constants depending on the circuit connectivity. Each kind of term in (3) has a simple physical interpretation: When the connectivity of the circuit is such that current flows through transistor j in order to charge a capacitance associated with transistor i, then $a_{ij}$ is non-zero and the first kind of term represents an RC-delay associated with this resistance charging this capacitance. This RC-delay varies proportionately with $x_i$ and inversely proportionately with $x_j$. A constant amount of time is required for a transistor to charge its own drain. This time is given by terms of the first kind with i = j. The second kind of term represents the RC-delay associated with a transistor charging a constant capacitance such as an off-chip or wire capacitance. As was noted in [Fish85], formula (3) is a *posynomial* in the $x_i$, and the sum of transistor sizes A is also a posynomial.

## Transistor Sizing Problem is a Posynomial Program

We can now see that the problem

"Minimize A subject to the constraint T<K"

is the minimization of one posynomial (A) subject to a set of upper-bound constraints on posynomials (delay along each path must be less than K). In other words, the transistor sizing problem as we have formulated it is a *posynomial program* [Ecke80]. When viewed as a function of the logarithms of its variables, a posynomial is *convex*, which means that any straight line segment in N+1-dimensional space whose endpoints lie in the graph of the function is itself entirely on or above the graph. With this transformation, it can be seen that a posynomial program is a special case of a *convex program* and thus enjoys the special properties of convex programs. The most important of these is that a local minimum is guaranteed to be a global minimum. Thus we need not worry about one of the major headaches of VLSI CAD, *viz.* entrapment in local minima. Computationally expensive techniques that have been devised to escape from local minima, such as simulated annealing, multiple random starts, or the Kernighan-Lin heuristics [Kern71] for min-cut graph partitioning, are unnecessary.

In [Fish85], it was incorrectly stated* that the posynomials that describe the delay along a circuit path were convex in the transistor sizes themselves. This error was the result of a back-of-the-envelope proof that was perfectly correct except for the final step, from the next-to-the-last statement in the proof to the (false) theorem statement. Fortunately, due to the above transformation, the main result of this false theorem is still true: A posynomial function can be viewed (by using "log scale" coordinates for the independent variables) as a convex function (of the logarithms of the variables). Thus a posynomial program can be viewed as a convex program, and the field of posynomial programming is a subfield of the field of convex programming. In a posynomial program, a local minimum is always a global minimum.

---

* The error was pointed out to us by D. P. Marple [Marp86-1]. We replied by pointing to the geometric programming literature [Ecke80, Pete76], which details the transformation from posynomial to convex program. It was then agreed by all that "Minimize A subject to the constraint T<K" for static CMOS circuits with simple-RC delays was indeed equivalent to a convex program, and thus had the local-optimum-is-global-optimum property. Marple later put posynomial programming to good use in his PhD thesis.

We will call RC timing models "simple-RC" if they do not take into account the slope of the gate input waveform, "slope-RC" if they do. The above posynomial model applies to static CMOS and precharged logic as long as the simple-RC timing model is used. This model does not apply to ratioed-load gates [Mead80], no matter whether simple-RC or slope-RC timing models are used. This is due to the fact that when the pulldown network is turned on, the current of the pullup resistor must be subtracted to obtain the output current. The above posynomial model also does not apply to any logic family when input slope of the gate input signal is taken into account in the delay model. This is due to the fact that the only closed-form function [Horo84] that is available for slope-RC delays is much more complicated than the rather simple formula for simple-RC delays. Recently Marple [Marp86-2] showed that with suitable transformations, the problem of minimizing the area of a precharged PLA to meet given timing constraints can be cast as a posynomial program, even when using the slope-RC delay model. Presumably these transformations could be applied to other kinds of precharge logic, such as domino or zipper, as long as some suitable approximation to cost is used, such as sum-of-transistor-sizes.

To summarize the theoretical side of the transistor size optimization state-of-the-art: The only circuit types that have been proven to have the local-optimum-is-global-optimum property are the two that have been cast as posynomial programs:

1. Static CMOS with simple-RC delay model [Fish85].

2. Precharged logic gates with slope-RC delay model [Marp86-2].

For all other circuit types, it is not currently known how to formulate the optimization problem so that the local-optimum-is-global-optimum property can be proved. These include ratioed-load logic (nMOS and CMOS "pseudo-nMOS"), static CMOS with the slope-RC model, transistor-transistor logic (TTL), and emitter-coupled logic (ECL).

**How Transistor Sizing Really Helps**

All this theoretical mumbo-jumbo is perhaps satisfying to a mathematician, but we owe the circuit designer an explanation of what sizing does to a circuit to make it go faster. We will give this explanation in the form of an answer to the question:

WHY ARE RANDOM LOGIC GATE DELAYS LARGER
THAN RING OSCILLATOR GATE DELAYS?

Assume for the moment that all transistors are a single, fixed size. In both random logic and a ring oscillator, the delay of a gate is an RC delay. Roughly speaking, the resistance R is proportional to the *fanin* of the gate. The capacitance C comes

from two sources: the capacitance of the fanout transistor gates, and wire
capacitance. And now we have the answer to our question: A random logic gate
delay is in general larger than a ring oscillator because

1. The fanin of a ring oscillator gate is one, whereas the fanin of a random
   logic gate is greater than one.

2. The fanout of a ring oscillator gate is one, whereas the fanout of a random
   logic gate is greater than one.

3. The wire capacitance of a ring oscillator gate output is negligible, whereas
   the wire capacitance of a random logic gate output is non-negligible.

What does this have to with transistor sizing?

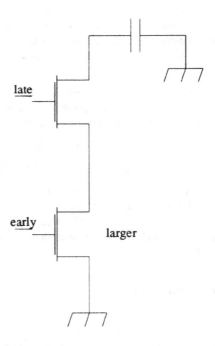

Figure 7-4: Transistor Sizing Used To Reduce Delay Due To Fanin.

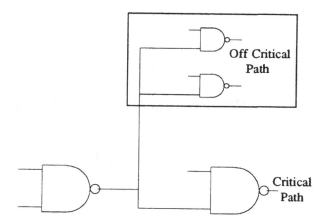

Figure 7-5: Transistor Sizing Used To Reduce Delay Due To Fanout.

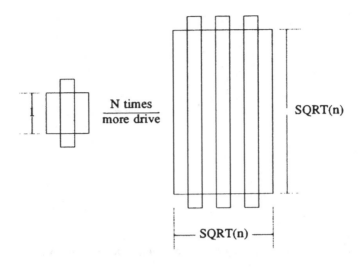

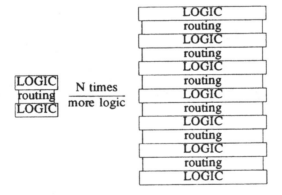

Figure 7-6: Transistor Sizing Used To Defeat Wire Capacitance.

1.  **Transistor sizing can help to reduce the delay due to fanin.** When two
    transistors are in series, the one whose input arrives earlier can be made
    larger and attached closer to the power rail (Figure 7-4). When the later
    signal finally arrives, the output can be driven with resistance not much
    more than that of the top transistor.

2.  **Transistor sizing can help to reduce the delay due to fanout.** When a
    gate drives several other gates, only one of which is on the critical path, the

transistors in the gates off the critical path can be made smaller (Figure 7-5). This reduces capacitance in the driven node, thus speeding up the critical path at the expense of the other paths.

3. **Transistor sizing can be used to defeat wire capacitance.** Suppose every transistor in a Gate-Matrix style layout were to have its drive increased by a factor of $N$ by increasing its channel width by $\sqrt{N}$, and by replicating it in parallel $\sqrt{N}$ times (Figure 7-6). The worst this could do to the layout would be to increase the height and width of each logic row by $\sqrt{N}$. The worst this could do to any wire on the chip would be to stretch it by the factor $\sqrt{N}$. Thus the wire capacitances are increasing at the rate $\sqrt{N}$, but the device drive and the device capacitances increase at the rate $N$. By making $N$ arbitrarily large, the extra delay due to wire capacitance can be made arbitrarily small.

It is important to note that this last argument assumes that only wire *capacitance* is a problem. Delay due to wire *resistance* is a far more serious problem and one that cannot be overcome directly with transistor sizing. Multi-layer metal capability has helped, but we are approaching the day when feature sizes will be so small and chip sizes so large, that the DC resistance of aluminum will contribute significantly to delay.

## The TILOS Heuristic

Since we are interested in applying the sizing algorithm to circuits as large as an entire VLSI chip or perhaps to an entire system, the algorithm must be as efficient as possible, perhaps at the expense of absolute accuracy in finding the optimum. In our experience, the algorithm described below provides an efficient method for approximating to the optimum point. It was our decision not to use any of the non-linear optimization techniques, since these have generally used too much time and memory for problems of this size. The TILOS heuristic, on the other hand, has been used successfully on circuits with more than 40,000 transistors.

TILOS proceeds as follows: Starting with minimum sizes for all transistors* a static timing analysis is performed on the circuit, which assigns two numbers to each electrical node: $t_l$ (latest time to go low), and $t_h$ (latest time to go high). TILOS determines the single primary circuit output or latch input that is failing by the greatest amount to meet its timing goal for $t_l$ or $t_h$. Let us call this node N.

---

* The user can specify the starting sizes of transistors individually, or request that sizes of specified transistors be frozen and hence not subject to optimization. These features are important when interfacing with parts of the chip not processed by TILOS.

TILOS walks backward along the critical path leading to N. Whenever a node X is visited, TILOS examines in turn each nFET (if X's $t_l$ is failing) or pFET (if X's $t_h$ is failing) which could have an affect on the path. In general, this includes both the *critical transistor* (the transistor whose gate is on the critical path), the *supporting transistors* (transistors along the highest resistance path from the source of the critical transistor to the power supply), and *blocking transistors* (transistors along the highest resistance path from the drain of the critical transistor to the logic gate output). Using equation (3), TILOS calculates the *sensitivity* of each such transistor i, which is the time savings accruing per increment of $x_i$. The size of the transistor with the largest sensitivity is increased by multiplying it by the constant BUMPSIZE, a user-settable parameter that defaults to 1.5. The static timing analysis is incrementally updated and the entire process is repeated. The sizing process terminates when either the constraints are met or when the circuit has passed its absolute minimum and is getting slower instead of faster.

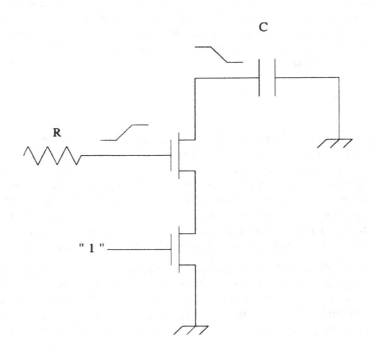

Figure 7-7: Calculating Sensitivities.

Figure 7-7 illustrates a series configuration in which the critical path extends back along the gate of the top (critical) transistor. The sensitivity for this transistor is calculated as follows: Fix all transistor sizes except x, the size of the critical transistor. R is the total resistance of an RC chain driving the gate. C is the total capacitance of an RC chain being driven by the configuration. Then the total delay D(x) of the critical path is

$$D(x) = K + RC_u x + \frac{R_u C}{x},$$

where $R_u$ and $C_u$ are resistance and capacitance of a unit-sized FET. K is a constant that depends on the resistance of the bottom transistor, capacitance in the driving RC chain and resistance in the driven RC chain. The sensitivity $D'(x)$ is then

$$D'(x) = RC_u - \frac{R_u C}{x^2}.$$

The sensitivity calculation for supporting transistors is done in a similar way. When $D'(x)$ is set equal to zero the resulting value of x, which minimizes delay, is equal to a constant times

$$\sqrt{C/R}. \tag{4}$$

An interesting consequence of (4) occurs in the special case of a string of inverters. Since C is proportional to the size of the next inverter in the chain and 1/R is proportional to the size of previous inverter, (4) tells us that, to minimize delay, the optimal size of the inverter is the *geometric mean* of the sizes of the next and previous inverters. In other words, the size ratio of consecutive inverters should be a constant. Of course we already know that the optimal size ratio is approximately e=2.71828..., so if we found a string of inverters with less than this ratio, we might be well advised to remove two inverters and try again.

Apparently (4) is true independently of the relative sizes of the first and last inverters, which leads us to the interesting, but seemingly useless, observation that in the case when the first inverter is *larger* than the last, the minimum propagation delay is achieved by geometrically *decreasing* sizes. But a commonly committed mistake in input buffer design shows us that this is not a useless observation after all: One often sees chip input buffers in which the input signal, after coming on chip and going through a lightning-arrester network, is fed into an inverter of minimum or near-minimum size. The output of this inverter is then fed into a string of geometrically-increasing inverters to drive the on-chip fanout. But to minimize delay we need to expand the picture to include the inverter on some other chip that actually drives the input pad. This inverter is necessarily large, to drive the large capacitance of the circuitboard wire. At the other end of the inverter chain, the on-chip fanout is small by comparison. For minimum delay, the inverters in between, in the input buffer, should be geometrically decreasing in size. And here too, removing pairs of inverters from the chain is desirable: the fastest input buffer is simply a wire from the lightning arrester to the on-chip fanout. (If the input signal is at TTL levels, the TTL-to-CMOS conversion would then have to be performed at each of these fanout gates.)

## Some Practical Results

TILOS has been used to optimize parts of many experimental and production CMOS chips, including the Prefetch Program Counter of the WE$^{TM}$ 32200 microprocessor [Huan87]. In order to physically measure the speedup achievable with transistor sizing, several 32-bit adders were fabricated on a 1.25μ CMOS chip [Shug86]. Each adder consisted of a ripple-carry chain of 32 one-bit full-adders. The full-adder contained two complex gates implementing the logic equations

$$ZCN = ! ((A + B)C + AB)$$

$$ZSN = ! ((A + B + C)ZCN + ABC)$$

plus one inverter to generate the non-inverted sum and a second inverter to generate the non-inverted carry-in for the next full-adder. In order to measure the delay along the carry chain, each adder was made to oscillate like a ring oscillator by feeding the final carry-out back into the carry-in through an inverter. The oscillation could be turned on or off by feeding a suitable pattern of 1's and 0's into the 64 inputs of the adder. One of the adders was implemented with standard cells from AT&T's High-Speed Polycell library. A second adder used the Area-Efficient Polycell library. The only difference between these two libraries is the transistor sizes. The pFET/nFET sizes for the High-Speed and Area-Efficient libraries are 28/20 and 5/5μ, respectively. Eleven more adders, corresponding to eleven different delay requests, were generated with TILOS/SC2 using the same transistor schematic as the standard cell adders. TILOS's sorting of series-connected subnetworks was not turned off, but it did not change any of the carry-chain gates, since the polycell schematic was already in the optimal configuration.

All 13 adders were fabricated on a single chip with a 1.25μ CMOS process, along with a conventional ring oscillator. The average gate delay and silicon area for these 13 adders are plotted in Figure 7-8. The conventional ring oscillator delay is represented by the dotted line. We wish to make several points about these data:

1.  The TILOS/SC2 circuits were compacted with a virtual-grid compacter. With a state-of-the-art compacter (such as MACS), one would expect the TILOS/SC2 curve to move downward 20% to 30%, as well as slightly to the left (due to decreased wire capacitance).

2.  Since the adder used a ripple-carry chain, the contribution of wire capacitance to propagation delay along the carry chain was negligible for all 13 circuits. Thus, the obtained speedup was due entirely to overcoming the effects of fanin and fanout. When wire capacitance is larger, and when

TILOS can be told the per-wire capacitances, one can expect even more speedup from transistor sizing.

3. The carry chain, which compromises a significant fraction of the adder, is the "critical path" of the circuit. When a smaller fraction of the circuit is critical, the benefits of transistor sizing are increased, since transistors in the non-critical portions of the circuit can be made smaller.

4. The two points representing the standard cell circuits lie above and to the right of the TILOS/SC2 curve (the Area-Efficient circuit just barely so). Thus for this circuit and for these two criteria, propagation delay and silicon area, the capability of TILOS/SC2 can be judged to be uniformly superior to that of the two Polycell Libraries.

5. The delays were calculated from oscilloscope measurements from a single chip of the periods of oscillation of the 13 adders. Thus there are no errors due to simulation model or process variation.

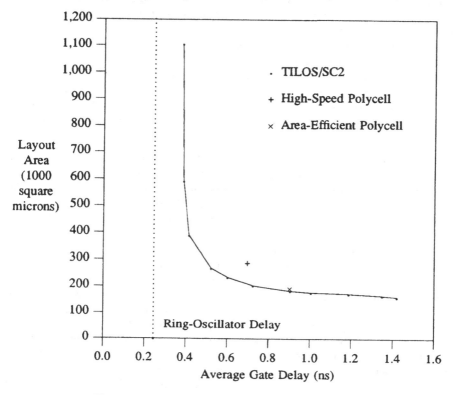

Figure 7-8: TILOS/SC2 Versus Standard Cell.

## Things That Still Need Fixing

While TILOS has proven effective and efficient at tuning large circuits, there is still room for enhancement. One problem is the underlying model of delay. The major source of error in the distributed RC timing model is the lack of consideration for slowly rising inputs. Unfortunately, except for precharged PLAs [Marp86-2], it has not been possible to prove for the slope-RC model, as it was for the distributed simple-RC model, that the transistor sizing problem can be cast as a convex program. Although there are several static timing analyzers and a transistor sizer [Mats85] that take into account input waveform shape for CMOS circuits, we hesitate to do so without a convexity proof in hand. If a more accurate model turns out to be non-convex there is always the danger that the optimizer might become trapped in a local minimum that is not a global minimum, resulting in a worse solution than the less accurate model.

Another source of error comes from the use of A, the sum of transistor sizes, as the measure of the circuit's area. In the final layout it is generally the case that some transistors, such as minimum-size ones, can be made larger without increasing overall area. Thus A is not really an accurate gauge of layout area. When using a particular layout style, such as PLAs, the actual layout area can be explicitly and accurately represented in the transistor-sizing program [Marp86-2]. For a more general layout style that is more amenable to variable transistor sizes, such as Gate-Matrix, it is not clear how to change the objective function without generating new local minima in the optimization program. Progress in this direction, however, is reported in [Ober88].

We have argued that the solution given by TILOS's heuristic is a good approximation to the optimum point of the program "Minimize A subject to the constraint T<K." Some experimental verification of this contention is available; in Table III of [Shyu88], TILOS came within 5% of the globally minimum A value in 8 out of 10 sample circuits. TILOS can perform poorly, however, when the circuit contains large fanin or fanout gates, or when the user requests that circuit performance be pushed to the limit. Although algorithms are available for converging exactly to the optimum point, none are currently as fast as TILOS or able to handle circuits with tens of thousands of transistors. One would hope that eventually non-linear optimization methods could produce an exact algorithm that can optimize large circuits at least as fast as TILOS. Significant progress is also being made on this problem; see for example [Shyu88]. In a sense, Shyu's method combines the best of the heuristic and exact methods, using the TILOS heuristic to quickly enter the feasible region, and then converging to the optimum point (without ever leaving the feasible region) with an exact method. If the user desires precision, this can be achieved. If the problem is so large as to require too much time by the exact method the user can set a "time-out" alarm. When the alarm goes off, the program would report its current state. Due to the fast method

for entering the feasible region, this will generally be a point meeting the timing specifications.

One final problem consists of the restrictions that TILOS places on circuits that it will accept. As a sampler, TILOS will not currently accept precharged or ratioed-load gates, latches made from cross-coupled logic gates, or clock-generation circuitry. Each restriction on circuit-type both simplifies the tool and inconveniences those users who want to use the outlawed style. Decisions as to what to allow and not allow were made largely according to the author's own prejudices and estimations of software complexity. We feel that as the level of integration available in CMOS continues to increase, these decisions will wear well: The user will reap the benefits of the new technologies by having a tool that can handle large designs, yet still deliver the higher speeds of the shorter channel lengths.

## 7.2 SLICC - An Alternative to Schematics

The previous chapter and this one have discussed tools that automate the tuning and layout of switch-level models of VLSI chips. For a designer, they raise the issue: Where do you get the models? One way is through conventional schematic capture systems, including the icon editor when used in this way. While a good schematic can be a thing of beauty, they are relatively slow to generate and tedious to update in the face of design changes. Another is through logic optimization and gate level synthesis tools, such as MIS [Bray84]. The output of such tools is an abstract multi-level logic format. If desired, this can be run through a technology binding tools such as DAGON[Keut87] and bound to a finite set of standard cells. This model can be converted to a transistor model by library lookup. But this model does not take full advantage of the flexibility of TILOS/SC2D, because it is constrained to a finite set of gates, not the potentially unlimited number possible by transistor synthesis.

What is needed is a tool that can generate an unconstrained set of netlists quickly and painlessly. Such a tool must support the specification of both the interconnection of logic cells, and the transistors that comprise these cells. In particular, it should take advantage of the additional freedom possible in an environment not limited to a fixed set of cells.

The approach taken by SLICC is not to invent yet another language for describing circuits at the switch-level, but rather to make use of an existing language, "C." C's syntax has been used verbatim, and wherever possible C semantics have been preserved. SLICC translates a dialect of C directly into the hierarchical transistor connectivity language "simnet." SLICC does not restrict the output to PLAs or a library of predefined cells. In fact, SLICC would be an appropriate tool for specifying and building such a library.

**Converting Syntax to Transistors**

The algorithm SLICC uses is based on electrical properities of MOSFETs. Generating circuitry for combinational logic is done by direct application of these principles to a parse-tree representation of the source expression. For example, the statement

    x = !(a && b)

would be translated using Algorithm 7-1:

1.  The expression would be tested to see if it is in an And-Or-Invert (AOI) format (*e.g.* only non-inverted variables inside an inversion). If not, all inverted subexpressions variables would be replaced with fresh variables.

2.  The subtree "a&&b" would be processed one literal at a time, yielding one nFET whose gate is connected to "a" one nFET with gate "b." A new net is allocated for the interior node. One end of the series circuit is connected to Vss and the other to "x."

3.  The pFET side of the tree is generated with two transistors wired in parallel. One side is connected to Vdd and the other to "x."

Algorithm 7-1: Translating Syntax to FETs

The result is shown in Figure 7-9a.

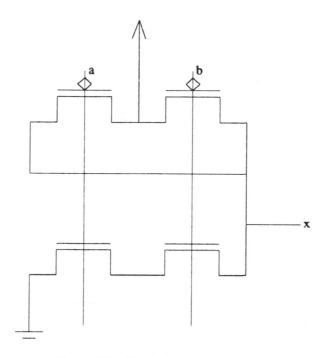

Figure 7-9a: Result for x = !(a && b).

An example of a non-AOI expression would be:

     y = e && !(f ‖ g);

This expression would be handled by first converting it to the form

     y = !(!(e && !(f ‖ g));

New literals, "zz0" and "zz1" would then be defined as

     zz0 = !(f ‖ g);
     zz1 = !(e && zz0);

and finally "y" would be redefined as

     y = !zz1;

At this point, the circuit for "zz0," "zz1," and "y" can all be generated directly. The result is shown in Figure 7-9b.

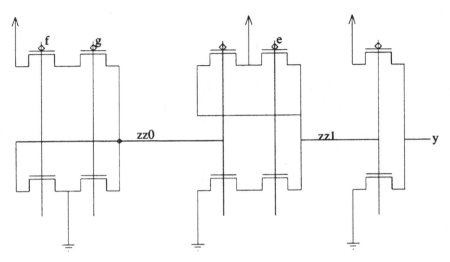

Figure 7-9b: Result for    y = !(!(e && !(f || g)).

In addition, SLICC understands DeMorgan's Law, and so recognizes that

  y = !b && (c || !d) ;

can be efficiently implemented with two gates as in:


zz1 = !c;
y = !(b || (zz1 && d));



**Eliminating Constants**


SLICC will simplify any logic expression that involves translation-time constants. Thus if A and B are defined to be 0 and 1, the expression:

  z = (!(y && !A)) || !B;

turns into

  z =  !y;

## Eliminating Redundant Gates

The above translation has the disadvantage that it can introduce redundant circuitry where none existed on input. For example, the expressions

    a = e && !(f ‖ g);
    b = w && r && !(f ‖ g);

would both generate a gate to compute "!(f ‖ g)." To eliminate this overhead, SLICC examines the set of hardware gates generated and removes redundant ones. The algorithm to do this is given in 7-2.

1.  Compute a hash function for each gate based on its inputs and logic function.

2.  Compare logic cells that have the same hash value.

3.  Eliminate redundant entries, and rewrite the circuit to make use of only the non-redundant gate outputs.

Algorithm 7-2: Eliminating Redundant Logic Gates

The algorithms described here are comparatively simple algebraic techniques - that is, they do not take advantage of the deeper factoring methods used in tools like MIS. However, they have the advantage of running quickly and always producing logically correct circuitry. Other, more sophisticated techniques exist for generating more nearly optimal transistor designs [Lai85]. However, in practical designs, it is not clear that these optimizations should be done at the transistor level. This is because most designers working at this level often have special knowledge of their application, and often want their circuits to be sub-optimal in the conventional metrics (such as number of transistors) but better for their particular needs. Such considerations include circuits where some input is known to be late settling, and hence the circuit must be rearranged to put that input closer to the output. In other cases redundant logic values must be computed because they will be needed at geometrically distant points. This is easy to accomplish with SLICC, but difficult to express in a canonical logic-table format.

## Software versus Hardware Variables

To make SLICC more convenient, it supports the notion of *translation-time* variables, which do not generate any hardware directly. These can be used to condense and simplify the expression of complex logic since they enable the programmer to make use of C language constructs.

Many practical hardware systems exhibit regularity proportional to a word size. It would be tedious and error prone to require the user to refer to individual nets, so SLICC supports the C "[]" syntax for array references. These are expanded at translation time into names for the individual bits that the simulation and layout systems can handle.

A SLICC input program can be viewed as two programs in one. If no hardware is to be generated, then it is simply a C program. If hardware is to be generated, it must center around the nets (a.k.a. "signals") of the circuit. In SLICC, as in ESIM [Frey84] and CONES [Stro86], signals are declared with the predefined type "sig." Only the "sig" variables will be translated into hardware. The translation will be done under the control of the software program. For example, C loops can be used to describe regular or nearly-regular circuits. Arbitrarily complex expressions operating on the translation-time variables can be used to control the generation of logic. "If" statements in the input are handled in an unusual way: When the expression in them can be evaluated at translation time it is, and the logic inside the "if" statement is either included or excluded depending on the result. On the other hand if the expression contains hardware nets, they are mapped into "pass logic" [Mead80] in the output, something which is difficult to do in library or PLA-based systems. For example, the code:

```
#define WS 16
  sig mem_in[WS], mem_out[WS], mem_cycle;
  sig bus[WS];
  int i;
  for(i=0; i<WS;i++) {
    /* generate gated logic */
    if (mem_cycle)
        bus[i] = mem_out[i];
    /* translation-time "if" */
    if (i < WS/2) {
    mem_in[i] = bus[i+8];
      };
  };
```

would be mapped into circuitry that always drives "mem_in" from the high-order bits of "bus", and which drives "bus", but only when the net "mem_cycle" is true.

## Exclusive OR and Decoders

The C operators "!=" and "==" are interpreted as exclusive OR and exclusive NOR respectively. Since the exclusive OR of a hardware variable with a constant is either the true or complemented form of the variable, a decoder can be described tersely in SLICC, as in

```
sig in[4], out_b[16];
int i, bit0, bit1, bit2, bit3;
for(i=0; i<16; i++) {
   bit0=(i&1)==1; bit1=(i&2)==1;
   bit2=(i&4)==1; bit2=(i&8)==1;
   out_b[i]=!(in[0]==bit0 && in[1]==bit1
     && in[2]==bit2 && in[3]==bit3);
   };
```

This generates 16 NAND gates that decode the 4 bits of "in."

## Lower-Level Specifications

Some circuits cannot be expressed as assignments and "if" statements. To support these, SLICC supplies the function "fetpath", as in:

```
fetpath(f, t, p, polarity);
```

which constructs transistors running from the net "f" to the net "t" and passing through the path "p." Path "p" must look like a C expression. "Polarity" may take on the values "nmos", "pmos" or "dual". Thus

```
fetpath(a,b,c,"nmos");
```

would join "a" to "b" with an nFET whose gate is "c."

```
fetpath(x,y,(c &&d) || e, "nmos");
```

would join "x" to "y" with three transistors, these being "c" and "d" wired in series, and "e" wired in parallel. The "dual" option is equivalent to the "nmos" option followed by the "pmos" option on the dual of the expression.

## Geometric Placement

Designers working at the switch level often have a rough floorplan in mind. SLICC allows it to be expressed by including a position next to each statement using the C comma operator. For example,

```
a =! (b&&c), at(4,6);
```

will place a NAND gate at column 4 and row 6. These positions may include translation-time expressions, as in:

```
a[i]=!(b[i+1]&&c), at(col*2,i%4);
```

## Directives

SLICC supports directives to control its actions, these include:

1. Transistor dimensions: the size of transistors may be specified in absolute microns or a multiple of the default sizes for the underlying technology. The values may be translation-time expressions.

2. Technology: Currently, SLICC generates complementary CMOS gates, but could easily be taught to generate nMOS, pseudo nMOS, or with a little more effort DOMINO and ZIPPER logic.

3. Optimization strategy: Three directives are recognized: "speed," "area" and "verbatim." The first two influence the selection of circuitry when more than one possibility exists. For example, in CMOS it is possible to connect the output of an inverter into the input of a pass gate to form half a shift register. It is possible to optimize this circuitry for area by eliminating the internal connection from the p side to the n side. The result runs slower, but saves a contact width in the x dimension. Thus the code

```
directive("optimize","area");
if(phi1)
        a = !b;
```

results in the schematic shown in Figure 7-10a, and the code

```
directive("optimize","speed");
```

```
if(phil)
        a = !b;
```

results in the schematic shown in Figure 7-10b.

Directive "verbatim" suppresses the elimination of redundant gates. This is most useful then there are other considerations not known to SLICC, such as geometric placement, that necessitate the redundant generation of signals.

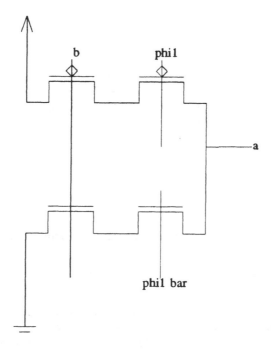

Figure 7-10a: Pass Logic with Directive Optimize Area.

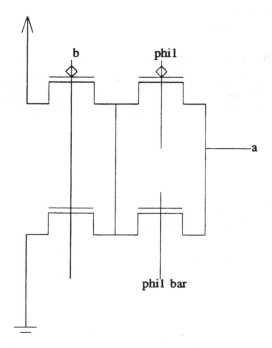

Figure 7-10b: Pass Logic with Directive Optimize Speed.

## Escape Mechanisms

Finally, an escape mechanism is provided to allow other directives to subsequent CAD tools to appear in the source code. These are useful because they allow the user to concentrate all available data into a single file. To help integrate the whole collection of switch-level tools, SLICC automatically generates a directory with its output files named according to standard conventions, including a UNIX "make" file that controls the switch-level sizer, simulator and layout tools.

## Libraries

SLICC can support libraries in several ways. For example, fixed libraries such as polycells can simply be referenced as functions with the same names. Layout generators such as a RAM generator can be programmed in the SLICC-style C as functions with software parameters. SLICC would then interpret these functions and generate a switch-level model for the configuration of RAM required. This would allow the chip to be simulated as a whole without running the generator and extracting its circuit geometrically.

Finally, the possibility exists for a whole new class of generators written specifically to take advantage of the SLICC/TILOS/SC2D environment. For example, an adder generator could be parameterized by the number of bits and the desired performance, and would emit a ripple carry, an inverted-pair ripple carry, a group carry or a carry look-ahead version depending on parameters. Writing such a generator, while quite a chore, still represents a small fraction of the effort required to write the type of geometric generator described in Chapter 6.

## Summary

The underlying assumption in SLICC is that today's designer of high-performance chips is skilled both in the techniques of programming and in the techniques of switch-level design, and capable of making his or her own design tradeoffs. SLICC represents a way to put that knowledge to use in the environment that he or she is familiar with, and for which a large collection of auxiliary tools is available. Furthermore, through the creation of libraries and switch-level generators, SLICC allows a single designer to multiply his/her impact without worrying about geometric details.

However, the biggest effect is not from the use of SLICC on its own, but rather as an adjunct to TILOS and other ways of generating circuits. The bulk of the logic in most chips does not require the detailed manipulations possible in SLICC. For these areas other tools such as CONES that more closely follow the semantics of C have advantages in terms of ease of use and better verification. But with both tools available, the chip builder can select from a wide range of possibilities in terms of speed of design versus speed of operation. Portions of a chip with loose timing considerations can be run through CONES and optimized with TILOS. Portions with tighter requirements can use predeclared SLICC libraries (possibly optimized with TILOS), and full-custom circuit design can be done directly in SLICC. All methods use the same input language, and can be simulated and synthesized together.

# Chapter 8
# Summary and Trends

This book has demonstrated how the principles listed in the introduction have been put to work in today's system. This chapter summarizes the key points, and provides some opinions on how the ideas can evolve into the future.

## 8.1 Summary

In the introduction of this book we stated several guiding principles of the IDA system. We evaluate the success of these principles by their impact on design methods, by their emigration into production tools, and by their acceptance in the external community.

- Constraint-based geometry: All of the tools read an write a common geometry-description language, IMAGES. Key to IMAGES is geometric constraints, which has proven to be unifying factor in automated layout tools. As described in Chapter 2, IMAGES translator solves a system of constraints based on design intentions. This relieves the author of IMAGES files (human or software) of a great deal of tedious design work, and allows floorplanning and routing information to be expressed naturally, as discussed in Chapter 5. At the cell level of design, such constraints are not heavily used *per se*, but they are used indirectly to interface with compacters. By constraining the ports on a symbol, pitch-matching and cell assembly is expedited. A similar, but somewhat simpler type of constraint resolution is used inside constraint-based compacters, such as MACS [Croc87]. These compact a layout by solving a system of constraints based on design rules. These would interact in constraint-based pitch-matching tools [Bull86], which solve a system of constraints based on a mixture of designer's intentions and design rules.

- Symbolic connectivity: Early geometric systems such as CIF did not tie together the notion of electrical nets with geometric objects. The IDA system does this through the notion of symbolic connectivity. Much of both the efficiency and complexity of geometric manipulations is due to the interaction between IMAGES and raw geometry. As discussed in Chapter 4, this

interaction is crucial to acceptance of the extractor, and makes the process of
design rule checking a meaningful comparison of design intent versus design
content.

- Technology independence and delayed technology binding: Given the rapid
  evolution of new processes, few today will argue against the advantages of
  technology independence. As discussed in Chapter 5, the IMAGES/IDA
  environment provides this in several ways depending on the needs of the
  particular chip. A small amount of technology updatability can be achieved by
  explicit references to technology constraints in the layout: a higher degree by
  extensive use of a compacter. The older style of generators provide some
  parameterization, including parameterization of technology, in fixed-geometry
  circuits such as shift-registers and adders. These have become common in
  production environments.

- Automatic layout synthesis: As discussed in Chapter 7, a higher level of
  technology updatability can be achieved by use of transistor-level design and
  transistor sizing, As discussed in Chapter 6, a newer style of general-purpose
  generators, such as SC2D, provides both independence from geometric
  technology rules and greatly reduced design time and effort.

- A common programming environment: Since the inception of the IDA system
  the tools have interacted with the UNIX computing environment. For example,
  the editor discussed in Chapter 3 uses UNIX pipes to interface with the
  simulator, and the SC2D system discussed in Chapter 6, uses UNIX facilities
  to control parallel execution within and across workstations. Today, UNIX has
  emerged as the standard for CAD systems. This has made it much easier to
  interface IDA software with externally provided tools.

- An additional degree of sharing and efficiency comes from the reuse of the
  internal IDA data representation, which was discussed in Chapter 2 and the
  Appendix. These structures have supported the interface with several new
  tools, thus providing a quick, and relatively efficient way to access and
  generate the IMAGES language.

Overall, the IDA/IMAGES environment has proven to be effective and useful in
today's environment. This is witnessed by the fact that new full custom chips are
built with it at the rate of about one per month. But what about the future? What
are the trends and how can they best be supported?

The IMAGES language and the IDA toolset are constantly being reevaluated and
tuned to meet the needs, current and anticipated, of today's designers and
tomorrow's chips. Key to this process is an understanding of the directions that
VLSI is heading and the motivations behind them. This chapter discusses some of
the trends in VLSI, and the way they influence CAD techniques.

## 8.2 VLSI: From Science Fiction to Commodity

Over the last ten years the idea of building a chip with 50,000 devices has changed from avant-guard to old hat. At the core of this change are dramatic improvements in lithography and other fabrication technologies. At the same time, other trends have emerged that have changed the shape of such designs and the way they are done.

### From Regular Geometry to Random Logic

Overall, the trend has been from highly regular geometric patterns to increased irregularity and more real-world complexity imposed on chips. Many early "academic" chips (such as the OM-1 [Mead80]) centered around a regular geometric pattern that tiled the surface of the chip, and the percent of the chip that was regular was a quoted metric. Such early chips often ignored the complicated issues of interfacing with an imperfect outside world. In most chips today, even RISC-based micro-processors, the highly regular geometry portion of the design represents a smaller and diminishing portion of the design work. This is a partly a consequence of increasing complexity in the overall design, and partly due to the improvements in CAD tools. But it is also due to the relative abundance of silicon.

### From Pushing Technology Limits to Run-of-the-mill Performance

It can be argued that silicon technology has advanced faster than the applications demanding it. In both time and complexity, the ratio of resources available to the demands of typical applications has increased dramatically. For example, consider the problem of manipulating video data that arrives at the rate of about 100 million 8-bit samples/second (a rate derived from national standards for broadcast television). In an older, nMOS technology, gate delays on the order of 2 nanoseconds were typical, which makes it impossible to add two samples with a simple, ripple-carry adder. So designers had to use more complex designs and cut margins to speed things up. Today's CMOS technologies have gate delays an order of magnitude faster, allowing the designer to use a simple adder and perhaps delegate the layout work to layout synthesis tools. Of course, researchers and advanced developers argue that the improved VLSI technology should motivate the adoption of higher sampling rates and more sophisticated processing algorithms. But the relationship has fundamentally changed: the consumers of raw computations are not searching for VLSI solutions so much as the VLSI suppliers are searching to increase the consumption of chips.

On the other side, in parallel with the increased resources for raw computing has come a new demand for better access to and control of the computing. The result is seen in the growth of the non-datapath portions of all chips, even chips that are data-intensive, like microprocessors, signal processing chips, and network control engines. The control portion has grown to dominate the chip design (if not the area). For example, typical microprocessors today (even RISC micros) have to deal with virtual memory, priorities and permissions, interrupt levels, pipeline breaks, floating point exceptions and so forth. Little, if any, of this is reflected in the regular portion of the registers, ALU, or cache. Most of it is reflected in the control logic that fills more than 50% of many such chips.

In early chips this logic was implemented in PLAs and microcode. But the trend in recent years has been to implement it in multi-level logic. Again the tendency is away from geometric regularity to random-appearing layouts. The motivation is speed: PLAs and microcode memories both imply a two-phase delay in most CMOS technologies, but multi-level logic does not impose any precharging/discharging delay. (In many shops, the PLA layout tools have become rusty from disuse in recent years.) More effort has been placed on multi-level optimization CAD tools such as MIS [Bray86] and multi-level technology binding tools such as DAGON[Keut87]. Since the original motivations for PLAs have largely disappeared, this trend seems likely to accelerate.

## 8.3 Design Style Trends

The change in the VLSI marketplace has necessitated a parallel change in the way chips are designed. Many chip projects strive not for the optimal, but to work within limits of design time, speed and area. Predictability is often more important to managers than maximum performance. This trend is motivated by several factors: First, chips are now viewed as just one component in a larger system. Often no single chip by itself determines the overall performance; the system as a whole has speed requirements and chips must simply work within them. Moreover, since devices have shrunk faster than packages, more chips are becoming pin limited and their cost is dominated by packaging. The result is that managers look to automated tools that will cut the variance from one chip design to another. They strive to make the design and layout process as predictable as possible, even if it means using a few more square microns of silicon. Such run-of-the-mill chips cannot depend on layout superstars, either in human form or in the form of special purpose generators, because the human may not be available at the critical time, and the special purpose generator may not produce exactly the right function. The trend for such chips is toward the general-purpose layout synthesis approaches: traditionally polycell, gate array, sea-of-gates, or the newer cell synthesis techniques like SC2D.

## CAD Tools as the Ultimate Authorities

On the subject of designer superstars, it seems that there are now tool superstars too. It used to be that the best designer's opinions were sought out on many critical issues. In many areas today ordinary CAD tools consistently produce better results than humans. Probably the first area that underwent this change was simulation. Circuit simulators, such as SPICE, are the most accurate way of evaluating circuit performance -- humans cannot solve the highly non-linear equations. Other evaluation areas are equally automated now: no human would claim to do a more reliable job at circuit extraction or design rule checking than the computers do. In synthesis, the trend is slower, but equally persistent. For large, random logic chips, today's computer systems for cell placement for polycell design, course routing (and maybe detailed routing), logic optimization, technology binding and transistor sizing are consistently better than even the best designers. And it should be noted that in each of these areas, the CAD tools are not mimicking humans as in "expert systems." The breakthroughs came instead through mathematical insight and solid algorithmic foundations.

## Compacters as Technology Binders

While not as widely accepted as other geometric tools, compacters are gaining acceptance throughout the industry. This is partly a response to the increased importance of design time relative to chip area. Even a simple compacter, such as a virtual-grid tool, represents a two way leap in designer productivity: First, it makes it much easier to lay out a given function because the designer does not have as many factors (such as design rules) to worry about. Secondly, it expedites the porting of an existing design to a new technology. As a bonus, it improves reuse within a technology because it provides parameterization of transistor sizes and pitch with little effort. These gains must be balanced against the overhead incurred learning the more sophisticated tools, and wasted area due to non-hand packing. But on balance, the gains outweigh the costs and future looks bright for compacters. Their acceptance should evolve in two ways:

First, one can expect to see improvements in the compacters themselves. For example, constraint-based compacters are gradually displacing virtual-grid compacters as "plug-compatible" replacements, enabling some circuits that would previously have required hand design to be done with compaction. Constraint-based compacters tend to provide better leaf-cell compaction, and have the property of producing better layouts with increasing sizes of compaction (unlike virtual-grid compacters that produce worse layouts with larger cells). Today, a state-of-the art compacter (MACS) can compact cells with 1000-2000 transistors in reasonable CPU and memory requirements. Within a few years, a combination of software and computing environment improvements should push that limit by an order of magnitude. Such abundant compacting resources are essential to make the full-custom circuit approaches like SC2D work: they depend on large amounts of compaction and storage to handle their "flat out" style of design. However, future compacter efforts will undoubtedly improve the hierarchical assembly process too. One tool for this purpose is PANDA [Bull86], a compacter-interface program that is integrated with the IDA design rule checker. PANDA works with compacters by calculating the minimum spacing among leaf-cells and routing modules.

The second trend in compaction is that good geometric compaction will obviate the need for other geometric CAD tools. For example, channel routers traditionally space the horizontal tracks apart by the distance required for two contacts to clear each other. A fancy router would sometimes go back and identify tracks where this did not occur, so that they could be spaced more closely. A very fancy router would add jogs to the route to allow wires from one track to fall into adjacent tracks [Chen88]. However, a jog-inserting compacter, working on the output of a naive router, can produce output of equal quality to a very fancy router indeed. Likewise, some early silicon cell synthesis tools worried about offsetting contacts to save area. Today's compacters can do a better job on this than the human designers, and this means that the cell synthesis tools don't need to worry about them. So the trend here is to push more of the geometric details into ever larger blocks and let the compacter fill in the details.

**Earlier Checkouts: From Geometric to Circuit**

Just as virtual-grid techniques abstract out tedious design rules, schematic level design abstracts out the geometric issues. In the halcyon days of VLSI, chips started out as pencil sketches, then progressed into layouts without being captured in machine readable form. The colored shapes acted as their own specification, documentation and realization. They were geometrically extracted, simulated and debugged, without benefit of an independent specification to compare them against.

In contrast to this is the design of the CRISP microprocessor that started out as a schematic, was debugged as a schematic, and when finally laid out was compared to the schematic without any further debugging or checking [Ditz86]. This movement from geometric checking to net-list checking is necessitated by the vastly increased logical complexity of designs, and by their geometric irregularity. For example, a highly regular circuit like a ripple adder is not much more complex geometrically than it is schematically. But a highly irregular piece of logic, like an exception controller or memory interface, is much more complex when represented geometrically than as a net-list. Because the geometry contains more information it is harder to understand, manipulate, and debug. The CAD consequence of this is that more tools are focusing on designing and tuning net lists, SLICC and TILOS being obvious examples.

Of course, schematic or net-level design is only one step on the road of increased design automation. Many designers today are not concerned with tuning the circuit parameters they deal with as much as they are with the logical behavior of the system as a whole. For that reason, gate-level tools such as MIS, when combined with technology binding systems like DAGON are attracting a growing following.

On the other side are chips that are not as complex logically, but which have demanding speed and/or power requirements. These chips tend to require all the support required for other chips, plus extra circuit level work, especially custom sizing of the transistors.

**The future: Automated Hand Design?**

The approach described in this book represents an evolution from hand design to synthesis. Many of the tools such as the editor, were originally intended for direct use by human designers. Although they are still used by some designers that way, they are less essential than they used to be. But even these have turned out to be helpful to automated synthesis subsystems. For example, the editor is used by SC2D via a command script for combining two strips of layout and routing into one larger cell before compaction. The compacter itself, which was originally intended to take human designs and convert them to masks, has turned out to be most useful as an assembler for layouts produced by SC2D and the routers. And at a more basic level the object-oriented, constraint-based language, IMAGES, which was originally designed mostly for human readability has proven to be a versatile medium for one tool to communicate with another.

# Appendix A
# Data Structures

## A.1 Basic IDA Data Structures

The IDA data structures have evolved over about five years to reflect thinking on the efficiency of storage, construction and access. Most of the structures were originally tuned to the needs of the icon editor, but they also reflect the needs of the IMAGES compiler.

## A.2 Basic Design Elements

The design is represented by a set of records* interconnected with pointers. The whole design is accessible from a globally accessible record called "world" which points to all the user-defined IMAGES symbols. Collections of objects are stored in rings. Rings are doubly linked lists with a single dummy element for a head. This relationship is illustrated in Figure A.1.

---

\* In "C", records are introduced by the keyword **struct**. This is unfortunate because there are many types of "structures" in software. To avoid potential confusion, and to avoid using non-English words wherever possible, we use the word *record* here.

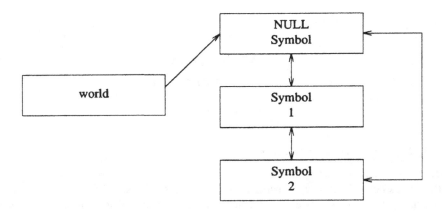

Figure A-1: Relationship between "World" and User Symbols.

The motivation for using rings, as opposed to simpler and more space efficient singly-linked lists, is that rings are slightly easier to manipulate. This is seen most clearly in the delete operation. In a singly-linked list one must provide some method for finding the preceding element in the list in order to update its pointer. In a doubly-linked list, one can remove any element independently. This is made even easier with the presence of the dummy header element which is never deleted. Without the dummy header element removing the last element in the set requires modifying the overriding record. But with the dummy element, removing the last element is exactly consistent with removing any other one. In the diagrams that follow, these rings are abbreviated by circles with double arrows in them, as shown in Figure A.2.

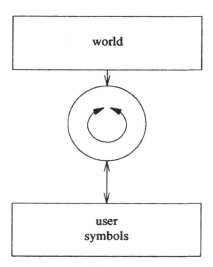

Figure A-2: The Ring Abbreviation of Figure A-1.

Each user-defined symbol points to rings of the objects within it. These include nets, ports, wires, devices, contacts, marks and blobs. This relationship is shown in Figure A.3.

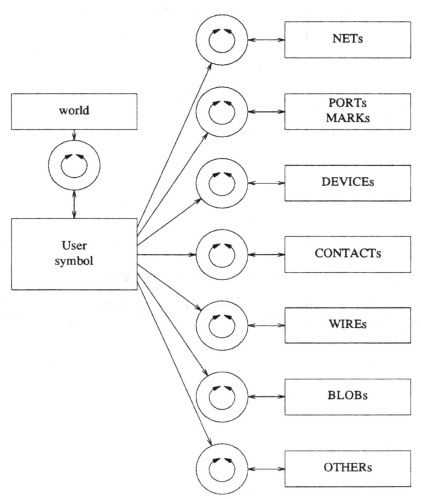

Figure A-3: Symbol and Its Objects.

193

The inner structure of a symbol is summarized here:

| name |
|:---:|
| PORTs and MARKs |
| NETs |
| WIREs |
| CONTACTs |
| DEVICEs |
| BLOBs |
| INSTANCEs |
| OTHERSs |

| mark-bit | check-bit |
|:---:|:---:|

The mark and check bits, which are available in all the IDA structures, are used to simplify the programming of the various algorithms that operate on them.

PORTs have a simple structure, as shown here:

| name |
|:---:|
| position |
| net |
| IMAGES-connection |
| level |
| name of global net |

| mark-bit | check-bit |
|:---:|:---:|

CONTACTs are similar to ports. NETs are slightly different, as shown here:

| name |
|:---:|
| number |
| connected net |
| inverted net list |
| value |

| mark-bit | check-bit |
|:---:|:---:|

The "connected-net" field in the net record is used for algorithms, such as union-find, that require a tree of nets to be constructed.

Wires are slightly different, since they have two independent endpoints. These are stored as an array internal to the wire record. Their structure is shown here:

| net |
| :---: |
| level |
| width |

| mark-bit | check-bit |
| :---: | :---: |

| end[0] |
| :---: |
| end[1] |

Each of the ends has a position and a pointer back to the wire record.

Devices are similar to wires, except that they have contact points (called device-ports) associated with them. These contact points contain a net, and a pointer back to the device structure enclosing them.

| name |
| :---: |
| device type |
| position |
| orientation |

| width | length |
| :---: | :---: |
| mark-bit | check-bit |

| device-ports |
| :---: |

Each of the device-ports contains a net and a pointer back to the device record.

Blobs are like wires, except that they contain a linked list of vertices.

| net |
|---|
| level |
| starting position |
| mark-bit | check-bit |
| pointer to first vertex |

The vertices contain a position, and a pointer back to the blob record.

Finally symbol instances are similar to devices, except that they contain a pointer to an array of ports that mirrors the ports in the defining symbol

| instance name |
|---|
| base symbol pointer |
| position |
| orientation |
| mark-bit | check-bit |
| pointer to instance-port array |

Each of the instance ports in the instance-port array contains a net, a pointer to the defining port, and a pointer back to the enclosing instance record.

For some tools, such as the graphical editor, a symbol is primarily a group of circuit elements, such as contacts and devices. A secondary quality of these elements is that they have constrainable geometric coordinates and electrical connectivity. Recall from Chapter Two that besides building the IDA data structures the principal enhancement of the IMAGES translator is that it resolves geometric constraints and electrical connectivity. To accomplish this the IMAGES translator relies on connections. An IMAGES connection is a symbolic coordinate in the three dimensional design space. A connection abstracts away the particulars of objects, such as contacts and devices, associated with a connection, but keeps two key features: the electrical net associated with the connection (for electrical connectivity checks) and the geometric constraints placed on the connection (for constraint resolution). A connection serves as a vertex in both the constraint graph associated with geometric constraints and the tree associated with electrical nets. It is only of secondary concern to the IMAGES translator that these connections are associated with circuit elements. This dualistic view of a symbol is shown in Figure A-4.

IMAGES connections are implemented in the connection data structure:

| legal layers |
| --- |
| net |
| coordinate[0] |
| coordinate[1] |
| constraints |
| invert connection |

The legal-layers field is a bitfield that gives the set of compatible layers that objects placed at the connection may have. Layer compatibility between connections can be checked by AND-ing the bitfields. The invert connection field points to a linked list of the objects associated with the connection. This allows one to trace back to the original user objects from the connections.

The constraints field of a connection points to the constraints record that contains all the geometric constraint information associated with a symbolic coordinate. The constraint information is associated with the constraint record:

| forward list[0] |
| --- |
| forward list[1] |
| backward list[0] |
| backward list[1] |
| count |
| in queue[0][0] |
| in queue[0][1] |
| in queue[1][0] |
| in queue[1][1] |
| fixed[0] |
| fixed[1] |

These fields are used in the rather complicated geometric constraint resolution algorithms described in Chapter Two. The forward-list and backward-list fields contain pointers to the forward and backward adjacency lists. Count is used to monitor the number of iterations in loops. The in-queue array is used to mark whether a vertex is in one of the four different queues. The fixed field tells whether the vertex has become absolutely constrained.

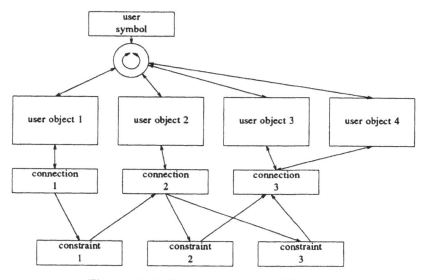

Figure A-4: Connections and Objects.

Finally, the system makes extensive use of sets of objects that are stored in rings called "groups". An example of this is shown in Figure A.5.

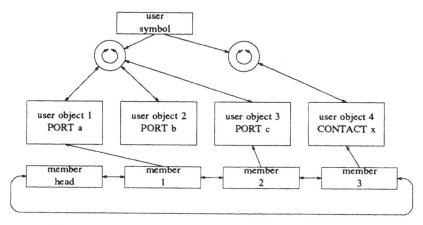

Figure A-5: A Group Pointing to Some Members.

# Appendix B
# Background on the "i" Language

The "i" language was invented by Steve Johnson to support the original Carver Mead design class at Murray Hill. Although it incorporated the same type of geometric constraints that were later used in IMAGES, it differed in a number of ways. Syntactically, it was slightly less verbose, in that it did not require a keyword in each statement the way that IMAGES does. For example, if "p" and "q" were XDEFs (called PORTs in IMAGES) then the statment

p RIGHT TO q;

would be equivalent to the IMAGES statement

WIRE AUTO p RIGHT TO q;

More significantly, the "i" language did not provide any support for the naming of electrical nets nor of assigning them properties. Nets numbers were stored internally, and electrical connectivity was derived in much the same way as IMAGES was to do later, but the nets did not have any names. Some of the early tools allow users to access them by referring to one of the physical objects associated with them.

In terms of large system design the "i" language had another serious deficiency: only global scoping of user symbols was provided. This meant that two independently generated designs could, and often would, conflict. In "i" it was solved through the use of the

FORGET <symname>;

operation. This was solved in IMAGES though the use of improved scoping techniques.

# References

[Ackl83] Ackland, B., N. Weste, "An Automatic Assembly Tool for Virtual-Grid Symbolic Layout," *Proceedings of VLSI 83*, 457-466, (1983).

[Agra82] Agrawal, V., "Synchronous Path Analysis in MOS Circuit Simulator," *Proc. 19th Design Automation Conf.*, pp. 629-635, 1982.

[AHU74] A. Aho, J. Hopcroft, J. Ullman, *The Design and Analysis of Computer Algorithms*, Addison Welsey, 1974.

[Bair78] Baird, H. S., "Fast Algorithms for LSI Artwork Analysis," *Journal of Design Automation and Fault-Tolerant Computing*, pp. 179-209, 1978.

[Bell58] R. Bellman, "On a routing problem", *Quart. Appl. Math*, 16, 87-90, (1958).

[Bona86] C. Bonapace, S. Pattanam, C-Y. Lo, "LARC2 Tutorial and Reference," *Technical Memorandum TM 52171-860918-01*, September 18, 1986.

[Bray84] Brayton, R, E. Detjens, S. Krishna, T. Ma, et al., "Multiple-Level Logic Optimization System," *Proceedings of the International Conference on Computer Aided Design*, October, 1984, pp 23-38.

[Bry80] Bryant, R. E., *MOSSIM: A Logic-Level Simulator for MOS LSI, User's Manual*, Integrated Circuit Memo 80-21, M.I.T. Department of EECS. 1980.

[Buri83] M. R. Buric, C. Christensen, and T. Matheson, "Plex -- Automatically Generated Microcomputer Layouts," *Proceedings of the 1983 IEEE International Conference on Computer-Aided Design*, Santa Clara, Ca, 1983.

[Bull86] Bullman, W. R., L. A. Davieau, H. S. Moscovitz, and G. D O'Donnell, "PANDA: A Module Assembler for the IDA Design Environment." *Bell Labs Internal Memorandum*, 1986.

[Chen88] Cheng, C-K, D. N. Deutsch, "Improved Channel Routing by Via Minimization and Shifting," *Proceedings 25th Design Automation Conference*, pp. 677-680, Anaheim, CA, 1988.

[Chu83] Chu, K-C., J. P. Fishburn, P. Honeyman, Y. E. Lien, "Vdd - A VLSI Design Database System," *Proceedings of the 1983 Annual Meeting -- Database Week*, IEEE Computer Society Press, 1983.

[Chu84] Chu, K-C., R. Sharma, "A Technology Independent MOS Multiplier

Generator,"" *Proceedings 21st Design Automation Conference*, Albuquerque, New Mexico, Jun 1984, pp. 90-97.

[Croc87] Crocker, W. H., Varadarajan, R., and Lo, C. Y., "MACS: a Module Assembly and Compaction System" *International Conference on Computer Design*, pp. 205-208, October, 1987.

[Ditz86] Ditzel, D, "Design of the CRISP Microprocessor", in *VLSI CAD Tools and Applications*, edited by W. Fitchner, Kluwer Academic Press, 1986.

[Dunl83] Dunlop, A. E., "Automatic Layout of Gate Arrays," *Proceedings of the IEEE International Symposium on Circuits and Systems*, Newport Beach, CA May 2-4, 1983, pp. 1245-1248.

[Ecke80] Ecker, J. G., "Geometric Programming: Methods, Computations and Applications," *SIAM Review*, Vol. 22, No. 3, pp. 338-362, July 1980.

[Fink74] Finkel, R. A., and Bentley, J. L., "Quad Trees: A Data Structure for Retrieval on Composite Keys", *Acta Informatica*, Vol. 4, pp. 1-9, 1974

[Fish85] Fishburn, J. P, and A. Dunlop, "TILOS: A Posynomial Programming Approach to Transistor Sizing," *Proceedings of IEEE International Conference on Computer-Aided Design-85*, Santa Clara, Ca. 1985.

[Fras78] Fraser, A. G., "Circuit Design Aides on UNIX," *SIGDA Newsletter* 1978.

[Frey84] Frey, E. J., "ESIM: A Functional-Level Simulation Tool," *Proceedings of ICCAD 1984*, 1984.

[Gee87] Gee, P, M. Y. Wu, S. M. Kang, and I. N. Hajj, "Metal-Metal Matrix (M cubed) CMOS Cell Generator with Compaction," *Proceedings International Conference on Computer Aided Design*, Santa Clara, November 1987, 184-187.

[Glas84] Glasser, L. A., and Hoyte, L. P. J., "Delay and Power Optimization in VLSI Circuits," *Proc. 21st Design Automation Conf.*, pp. 529-535, 1984.

[Gupt83] Gupta, A., "ACE: A Circuit Extractor," *Proceedings of the 20th Design Automation Conference*, pp 721-725, 1983.

[Hedl84] Hedlund, K. S., "Electrical Optimization of PLAs," *Proc. 22nd Design Automation Conf.*, pp. 681-687, June 1985.

[Hill83] Hill, D. D., "Edisim -- A Graphical Simulator Interface for LSI Design,"

*IEEE Transactions on Computer Aided Design of Integrated Circuits,* April 1983.

[Hill84-1] Hill, D., "ICON: A Tool for Design at Schematic, Virtual-Grid and Layout Levels," *IEEE Design and Test,* Vol. 1, 4, 53-61 (1984).

[Hill84-2] Hill, D., "CAD Systems for VLSI Design" *Proceedings of the National Communications Forum* Rosemont, September 1984, 673-691.

[Hill85-1] Hill, D., "SC2: A Hybrid Automatic Layout Program," *Proceedings International Conference on Computer Aided Design,* Santa Clara, November 1985, 172-174.

[Hill85-2] Hill, D. D. and S. Roy, "PROLOG in CMOS Circuit Design," *Spring COMPCON 85,* San Francisco, CA.

[Hill85-3] Hill, D. D., "SC2: A Hybrid Automatic Layout Tool", *International Conference on Computer Aided Design, 85* Santa Clara, CA., 1985.

[Hill85-4] Hill, D. D., "Effective Use of Virtual Grid Compaction in Macro-Model Generators" *Proceedings of the 22nd Design Automation Conference* Las Vegas, NV, 1985.

[Hill86-1] Hill, D. D., K. Keutzer, and W. Wolf, "Overview of IDA: A Toolset for Layout Synthesis," in, *VLSI CAD Tools and Applications,* edited by W. Fitchner, Kluwer Academic Press, 1986.

[Hill86-2] Hill, D. D., "Stretchable Routing in a Symbolic, Constraint-Based Environment," *International Conference on Computer Design,* Rye Town, NY 1986.

[Hill87-1] Hill, D. D. "SC2D: A Broad Spectrum Automatic Layout Tool," *Proceedings of the Custom Integrated Circuits Conference,* May, 1987.

[Hill87-2] Hill, D. "SLICC: A Switch-Level Synthesis System," *International Conference on Computer Design, 1987.*

[Horo84] Horowitz, M. A., "Timing Models for MOS Circuits," PhD thesis, Stanford University, January, 1984.

[Hsueh79] Hsueh, Min-Yu, *Symbolic Layout and Compaction of Integrated Circuits,* PhD thesis, University of California at Berkeley, December, 1979.

[Huan87] Huang, V. K. L., Altabet, S. K., Seery, J. W., and Wu, W. S., "The WE32200, AT&T's High Performance 32 Bit CMOS Microprocessor,"

*International Conference on Computer Design,* October, 1987.

[John77] Johnson, D., "Efficient Algorithms for Shortest Paths in Sparse Networks," *Journal of the ACM*, Vol. 24, 1, 1977.

[John82] Johnson, S., "Hierarchical Design Validation Based on Rectangles," *Proceedings Conference on Advanced Research in VLSI*, M.I.T., pp. 97-100, January 1982.

[Joup84] Jouppi, N., "Timing Analysis for nMOS VLSI," *Proc. 20th Design Automation Conf.*, pp. 411-418, June 1983.

[Kell81] K. H. Keller, "KIC: A Graphics Editor for Integrated Circuits," master's thesis, E.E. and C.S. Department, University of California, Berkeley, CA, 1981.

[Kern71] Kernighan, B.W and S. Lin, "An Efficient Heuristic Procedure for Partitioning Graphs," *Bell Sys. Tech. Journal*, Vol. 49 (2), pp. 291-308, 1970.

[Kern78] B. Kernighan and D. Ritchie, *The C Programming Language*, Prentice Hall, Englewood Cliffs, NJ, 1978

[Keut87] Keutzer, K., *DAGON: Technology Binding and Local Optimization by DAG Matching, Proc. 24th Design Automation Conf.*, pp. 341-347 June 1987.

[Koll84] Kollaritsch, P. W., and N.H. E. Weste, "A Rule-Based Symbolic Layout Expert," *VLSI Design*, August, 1984.

[Lai85] Lai, H. C., and S. Muroga, "Automated Logic Design of MOS Networks," *Advanced Research in Information Systems*, Vol. 9, 1985, Ed. J. T. Tou, Plenum Press, NY.

[Lee84] Lee, C. M., and Soukup H., "An Algorithm for CMOS Timing and Area Optimization," *IEEE J. of Solid-State Circuits*, Vol. 19, No. 5, pp. 781-787, Oct. 1984.

[Leng83] Lengauer, T., K. Mellhorn, "The HILL System: A Design Environment for the Hierarchical Specification, Compaction, and Simulation of Integrated Circuit Layouts," *Proceedings 1984 Conference on Advanced Research in VLSI*, M.I.T., pp. 139-147, 1983.

[Leng84] Lengauer, T., "On the Solution of Inequality Systems Relevant to IC-Layout," *Journal of Algorithms*, 5, pp. 408-421, 1984.

[LiWo83] Y-Z. Liao, C. Wong, "An Algorithm to Compact a VLSI Symbolic

Layout with Mixed Constraints", *IEEE Trans. on Computer Aided Design of Integrated Circuits and Systems,* CAD-2, 2, 62-69 (1983).

[Lin84] Lin, T.M., and Mead, C., "Signal Delay in General RC Networks with Application to Timing Simulation of Digital Integrated Circuits," *Proc., Conf. on Advanced Research in VLSI,* MIT, pp. 93-99, January 1984.

[LN82] Lipton, R., S. North, R. Sedgewick, J. Valdes, G. Vijayan, "ALI: A Procedural Language to Describe VLSI Circuits," *Proceedings 19th Design Automation Conference,* p. 467-474, 1983.

[LV83] R. Lipton, J. Valdes, G. Vijayan, S. North, R. Sedgewick, "VLSI Layout as Programming," *ACM Transactions on Programming Languages and Systems,* Vol 5, no. 3, 1983.

[Liao83] Liao, Y-Z., C. Wong, "An Algorithm to Compact a VLSI Symbolic Layout with Mixed Constraints," *IEEE Trans. on Computer Aided Design of Integrated Circuits and Systems,* CAD-2, 2, pp. 62-69, 1983.

[Lope80] Lopez, A.D., and H-F. Law, "A Dense Gate-Matrix Layout Method for MOS VLSI," *IEEE Transactions of Electron Devices,* Vol. ED-27, No 8, Aug 1980.

[Marp86-1] Marple, D. P., personal communication, August 18, 1986.

[Marp86-2] Marple, D. P., *Performance Optimization of Digital VLSI Circuits,* PhD thesis, Stanford University, October 1986.

[Math83] Matheson, T., M. Buric, C. Christensen, "Embedding Electrical and Geometric Constraints in Hierarchical Circuit-Layout Generators," *Proceedings International Conference on Computer Design,* p. 3, 1983.

[Mats85] Matson, M., "Optimization of Digital MOS VLSI Circuits," *Proc. Chapel Hill Conf. on VLSI,* U of N. Carolina, May 1985.

[Mead80] C. Mead and L. Conway, *Introduction to VLSI Systems,* Addison-Wesley Publishing Company, Menlo Park, Ca. 1980.

[Mell84] K. Mellhorn, *Graph Algorithms and NP-Completeness,* Springer-Verlag, 1984.

[Mont85] Montiero da Mata, J., "ALLENDE: A Procedural Language for the Hierarchical Specification of VLSI Layout," *Proceedings 22nd Design Automation Conference,* pp. 183-189, 1985.

[Moor57] E. F. Moore, "The shortest path through a maze," *Proceedings International Symposium on the Theory of Switching, Part II*, April, 1957.

[Nage80] Nagel, L. W., "ADVICE of Circuit Simulation," *IEEE Symp. on Computers and Systems*, 1980, Houston, TX.

[Nage75] Nagel, L. W., *SPICE2 - A Computer Program to Simulate Semiconductor Circuits,* University of California, Berkeley, ERL Memorandum Number ERL-M520, May, 1975.

[Ober88] Obermeier, F. W., and Katz, R. H., "An Electrical Optimizer that Considers Physical Layout," *Proceedings 25th Design Automation Conference*, p. 453-459, 1988.

[Oust81] Ousterhout, J. "Caesar: An Interactive Editor for VLSI Layouts," *VLSI Design*, pp. 34-38, Fourth Quarter 1981.

[Oust84] Ousterhout, J., "Switch-Level Delay Models for Digital MOS VLSI," *Proc. 21st Design Automation Conf.*, pp. 542-548, June 1984.

[Penf81] Penfield, P. and Rubinstein, J., "Signal Delay in RC Tree Networks," *Proc. of the 2nd Caltech VLSI Conference*, pp. 269-283, March 1981.

[Pete76] Peterson, E. L., "Geometric Programming," *SIAM Review*, Vol. 18, No. 1, January 1976.

[Reic86] Reichelt, M., *Improved Abstractions for Hierarchical Constraint-Graph Compaction,* Master's thesis, M.I.T., September, 1986.

[Rive82] Rivest, R.L., and C. M. Fiduccia, "A Greedy Channel Router," *Proceedings of the 19th Design Automation Conference*, Las Vegas, 1982.

[Rose84] Rosenberg, J., "Chip Assembly Techniques for Custom IC Design in a Symbolic Virtual-Grid Environment," *Proceedings Conference on Advanced Research in VLSI*, M.I.T., pp. 213-217, January, 1984.

[Rueh77] Ruehli, A. E., Wolff, P. K. Sr., and Goertzel G. "Analytical Power/Timing Optimization Technique for Digital System," *Proc. 14th Design Automation Conf.* pp. 142-146, June 1977.

[Sedg83] Sedgewick, Robert, *Algorithms*, Addison-Wesley, pp. 398-405, 1983.

[Shoj82] Shoji, M., "Electrical Design of BELLMAC-32A Microprocessor," *Proc. of the 1982 Int. Conf. on Circuits and Computers*, pp. 112-115 September 1982.

[Shug86] Shugard, D. D., and Dunlop, A. E., "Chip Layout Experiments using TILOS and SC2," *Bell Labs Internal Document* January 1986.

[Shyu88] Shyu, J. M., Sangiovanni-Vincentelli, A., Fishburn, J. P., and Dunlop, A. E., "Optimization-Based Transistor Sizing," *IEEE Journal of Solid-State Circuits* Vol. 23, No. 2, April 1988.

[Stro86] Stroud, C.E., R.R. Munoz, D. A. Pierce, "CONES: A System for Automated Synthesis of VLSI and Programmable Logic from Behavioral Models," *ICCAD*, Nov. 1986.

[Swar83] Swartz, P.A., B. R. Chawla, T. R. Luczejko, K. Mednick, and H. K. Gummel, "HCAP - A topological analysis program for IC mask artwork," *Proc. Intl. Conf. on Computer Design*, pp. 298-301, 1983.

[Szym84] Szymanski, T. G., and C. J. Van Wyk "GOALIE: A Space-Efficient System for VLSI Artwork Analysis," *IEEE Transactions CAD of ICs and Systems*, pp 278-280, 1984.

[Szym85] Szymanski, T. , "SOISIM User's Manual" *Bell Labs Internal Document* 1985.

[Szym86] Szymanski, T.G., "LEADOUT: A Static Timing Analyzer for MOS Circuits," *International Conference on Computer Aided Design*, Santa Clara, CA 1986.

[Tarj75] Tarjan, R. , "On the Efficiency of a Good but Non-Linear Set Union Algorithm," *Journal of the ACM*, Vol. 22, 2, (1975).

[Tarj81] R. Tarjan, "Shortest paths", *AT&T Bell Laboratories Technical Memorandum 81-11216-43* August 6, 1981.

[Tarj83] R. Tarjan, *Data Structures and Network Algorithms*, Society for Industrial and Applied Mathematics, 1983.

[Ullm84] Ullman, J. D., *Computational Aspects of VLSI*, Computer Science Press, Rockville MD, 1984.

[VHDL85] *VHDL Language Reference Manual, Version 7.2*, Intermetrics, Inc, Bethesda, MD IR-MD-045-2

[West81] Weste, N. , "Virtual Grid Symbolic Layout," *Proceedings of the 18th Design Automation Conference*, Nashville, Tenn., 1981.

[WIM86] Wimer, S., R. Pinter, J. Feldman, "Optimal Chaining of CMOS Transistors in a Functional Cell," *ICCAD*, Nov. 1986.

[Wolf85] Wolf, W., "An Experimental Comparison of 1-D Compaction Algorithms," *Chapel Hill Conference on VLSI*, Computer Science Press, Rockville MD, Henry Fuchs, May, 1985.

[Wolf86] Wolf, W., "Sticks Compaction and Assembly", *IEEE Design and Test*, Vol. 3, 3, 57-63 (1986).

# Index

active set, in scan-line 67
adjacency graph, SC2 147
algorithm for sizing 163
anti-features 102,74,92
attributes, in IMAGES 31
awk, use on IMAGES files 35
bounding box 96
branch-and-bound 72,74
breadth-first search 19
C language 169
CAESAR, editor 65
check, field 193,65
chosen group 39
chosing items in editor 39
comma operator, in SLICC 176
compacter, use in generators 121
compaction 185
compaction, virtual-grid 60
complex gate 157
CONES 174,179
connected, field 65
constraint resolution, in IMAGES 10
constraint resolution, use in routers
    110
constraints, cycles in graph 18
constraints, efficiency 29
constraints, equality constraints and
    union-find 12
constraints, in comparison with
    compaction 10
constraints, longest/shortest paths 10
constraints, non-positive edges 20,27
constraints, origin in IMAGES
    language 8
constraints, positive offsets 18
constraints, sets of equalities 12
constraints, topological ordering 20
convex program 158
CRISP 186

cursor, in editor 37
cycles, in constraints 18
cycles, in SC2 routing 143
DAGON 169
data structures, in IMAGES 189,65
data structures, in scan-line 66
databases, as alternative to IMAGES
    36
DeMorgan's Law 172
design rules, suppressing extra
    messages 100
design-rule checking 92
devices, in design rule checker
    100,101,95
devices, in net extraction 79
dog-leg routing 114
domino logic 159
doubly-linked list, ring 190
edges, in scan-line 66,95
editor, icon 37
efficiency of constraint solver 29
efficiency, in SC2D 149
electrical connectivity 30
eliminating constants and redundancy
    172
ESIM 174
exclusive-OR, in SLICC 175
exponential drivers, in TILOS 153
expressions, in IMAGES 32
extraction, fixed versus virtual 89
extractor, in editor 55
fanin 160
fanout 160
fetpath, in SLICC 175
field, check 193
field, mark 193
fillports operation, in editor 49
forest of trees 79
forget, "i" language 199

fuse operation, in editor 49
gate-matrix 75
generators 114
generators, IMAGES-C preprocessor 115
GOALIE 78
graphics interface 62
grep, use on IMAGES files 35
grow, geometric operation 70
HCAP 78
headers, in IMAGES 34
i language, drc 92
icon editor 189,37
IMAGES, constraint resolution 10
IMAGES, constraint-based design 9
IMAGES, deriving constraints from graphics 51
IMAGES, designing versus describing 5
IMAGES, enhancement of data structures 65
IMAGES, example: inverter 6
IMAGES, lexical analysis 32
IMAGES, loops 32
IMAGES, preprocessor 32
IMAGES, principal constructs 6
IMAGES, "connection" 195
IMAGES, "connection" as primary data structure 36
indirection array, in scan-line 67
intended nets 88
inverted net list, net extractor 88
inverter example 6
lambda 1
language, "i" 1
library, SLICC 179
loops, in IMAGES 32
make (UNIX utility) 3
make, in SC2D 138
make, use with generators 121
mark, field 193,65
mark, in editor 37
mates, in edge record 66
Mead, Carver 1,199

menu, in editor 38
metal2, in SC2 145
mincut, in SC2D 134
MIS 173
MOSFET model, in TILOS 155
mouse, in editor 37
move operation, in editor 48
multi-level logic 184
net extraction 77
net extraction, in editor 50
net naming, intended nets 88
nets, connected field 79
nets, naming 88
nets, nullnet 66
NFS (network file system) 150
notches, in design rule checker 102
nullnet, use of 66
optimization program 156
optimizing speed/area, in SLICC 177
parallel processing, in SC2D 149
patching anti-features, in design rule checker 102
patching notches, in design rule checker 102
pic 62
pitchmatching 105
placement, cost 131
placement, in SC2D 131
placement, manual 131
placement, random logic 134
placement, regular logic 132
port, data structure 193
posynomial program 158
posynomial, in TILOS 154
precharged PLA 159
pull-aparts 74
quad trees 72
ratioed-load gates 159
RC delay model, distributed 154
RC delay model, simple 158,159,168
RC delay model, slope 159,168
rectangles, merging 68
regularity, geometric design 183
resizing transistors in editor 40

reverse-polish editor commands 39
ring oscillator 159
ripple-carry chain 166
RISC 183
rise time considerations 168
routers 106
routers, channel 113
routers, dog-leg 114
routers, electrical specification 109
routers, floor plan specification 110
routers, ripple 113
routers, river 113
routers, using two passes 112
routers, via elimination 113
routing, in IMAGES 106
routing, in SC2 143
routing, in SC2D 134
SC2 129
SC2, metal2 use 145
SC2, routing 143
SC2, routing cycles 143
SC2, transistor flipping 141
SC2, transistor splitting 139
SC2D 129,154,169,179
scan-line 66,93
scan-line, in SC2 147
scan-line, indirection array 67
schematic/layout hybrids 52
scripts, editor 63
sensitivity 164
series connected subnetworks, in
    TILOS 155
sig variables, in SLICC 174
simnet 129
simulated annealing 134
simulation, batch 59
simulation, interactive 55
simulation, switch-level 58
SLICC 131,169,179
SLICC, comma operator 176
SLICC, fetpath 175
SLICC, library 179
SLICC, optimizing speed/area 177
SLICC, sig variables 174

SPICE 185
static timing analyzer 163
static timing analyzer, in TILOS 154
string of inverters 165
syntax tree 170
technology independence, IMAGES
    language 33
technology updatable design 121
TILOS 154,179
TILOS, distributed RC model 154
TILOS, exponential drivers 153
TILOS, MOSFET model 155
TILOS, posynomial function 154
TILOS, series connected subnetworks
    155
TILOS, static timing analyzer 154
TILOS, transistor sizing 153
top-down reading 33
top-down reading, editor 61
transistor flipping, SC2 141
transistor sizing, in TILOS 153
transistor splitting, SC2 139
translation-time variables 174
tub insertion 69
tubs 70
tubs, polarity conflicts 135
undo operation, in editor 48
union-find 79
union-find algorithm for connectivity
    31
union-find algorithm for placement 12
UNIV layer in IMAGES 54
UNIX, distributed 150
UNIX, forks 55
UNIX, pipes 55
variables, in IMAGES 32
vi text editor 38
via elimination, in SC2 147
via elimination, routers 113
via, in design rule checker 101
virtual-grid as schematic 5
wells, (see also tubs) 70
window, in design rule checker 101
wire capacitance 160

wire resistance 163
wire, data structure 194
X-windows 62